ALSO BY MARK SCHATZKER

Steak: One Man's Search for the World's Tastiest Piece of Beef

THE DORITO EFFECT

The Surprising New Truth About
Food and Flavor

MARK SCHATZKER

SIMON & SCHUSTER

New York London Toronto Sydney New Delhi

Simon & Schuster
1230 Avenue of the Americas
New York, NY 10020

First Simon & Schuster hardcover edition May 2015

SIMON & SCHUSTER and colophon are registered trademarks of Simon & Schuster, Inc.

For information about special discounts for bulk purchases, please contact Simon & Schuster Special Sales at 1-866-506-1949 or business@simonandschuster.com.

The Simon & Schuster Speakers Bureau can bring authors to your live event. For more information or to book an event, contact the Simon & Schuster Speakers Bureau at 1-866-248-3049 or visit our website at www.simonspeakers.com.

Manufactured in the United States of America

1   3   5   7   9   10   8   6   4   2

Library of Congress Cataloging-in-Publication Data

Schatzker, Mark.
The Dorito effect : the surprising new truth about food and flavor / Mark Schatzker.
pages cm
1. Junk food. 2. Nutrition. 3. Reducing diets. 4. Food portions. I. Title.
TX370.S33 2015
641.3—dc23
2014044543

ISBN 978-1-4767-2421-8
ISBN 978-1-4767-2424-9 (ebook)

For Laura

junk food *noun*

1. Pre-prepared or packaged food that has low nutritional value

   —*Oxford English Dictionary*

2. Food that is not good for your health because it contains high amounts of fat or sugar

   —*The Merriam-Webster Dictionary*

3. Food that tastes like something it is not

   —*Mark Schatzker*

Contents

PART ONE

THE DORITO EFFECT

ONE

"Things" and "Flavors"

IN THE early autumn of 1961, a thirty-seven-year-old housewife and mother named Jean Nidetch was pushing a shopping cart through a Long Island supermarket when she bumped into a woman she knew. "You look so marvelous," her friend said, and for a sweet moment Nidetch basked in the compliment. Unfortunately, her friend kept talking. "When are you due?"

Nidetch was not pregnant. At the time, she stood five seven and weighed 214 pounds, which marked her, in today's parlance, as obese, although Nidetch didn't know what that word meant, or that the obese were, at that very moment, coalescing into a demographic ripple that was on its way to becoming a wave.

Nidetch had been to see diet doctors in New York. When their advice didn't work, she headed across the Hudson River to New Jersey, where the diet doctors proved to be just as useless. She had tried every diet there was, and every one of them worked: She always lost weight. But then she would gain it all back—and more. Jean Nidetch could stop eating, just not for very long. She loved food too much. She loved savory things like pizza and meat, and sweet things, too, like cupcakes and soft drinks. Nidetch wasn't one for big breakfasts, but that was because she would get up at three in the morning to gorge on cold pork chops or baked beans right out of the fridge. In summer, if an ice cream, pizza, or sandwich truck zoomed by without stopping, she would take off after it. And when visions of jelly beans began dancing in her head, she would rifle through her son's pockets looking for some. But what Nidetch especially loved were cookies. When she started eating them, she couldn't stop. She was addicted to them.

The day Nidetch was mistaken for pregnant, she phoned the New York City Department of Health's obesity clinic to make an appointment. Not long after, she found herself in a room full of similarly overweight women. An instructor walked in who was so "slender" that Nidetch decided right there on the spot that after the class she was going to have an ice cream soda. The instructor handed out a sheet of paper with a list of foods the women were allowed to eat. Nidetch saw nothing new. She had whole albums filled with similar diets at home, none of which she'd ever been able to follow for very long. But once again, Nidetch tried. She gave up pizza, cake, and ice cream and started eating vegetables and fish. Every week, she went back to the obesity clinic, and every week she lost weight—two pounds.

It was progress, to Nidetch at least. The slender, ice-cream-soda-inducing instructor thought differently. She looked at Nidetch and said, "What are you doing wrong?" And as gallingly insensitive, perhaps even abusive, as that might sound, the instructor was right. The truth is Nidetch wasn't following orders, at least not completely. It was the cookies. She was feeding on them in secret. On the way to the clinic, she would sit there on the subway, constructing lies to explain her lack of weight loss, lies that got more and more elaborate with each passing week—I'm constipated, I'm retaining water, I'm premenstrual. By the tenth week, the shame had gotten so bad that she couldn't even look at the instructor.

Nidetch couldn't bear it any longer. She had to get her cookie secret off her chest, so she phoned six fat friends and invited them to her home and confessed. Her friends were supportive. She had a "right" to eat those cookies, they said. They did stuff like that all the time. One friend hid chocolate chip cookies in the cupboard behind dishes. Another hid snacks behind cans of asparagus where no one would see them. All of them confessed that they, too, got up in the middle of the night to eat. Toward the end of the meeting, something seemingly insignificant happened that would change the course of Nidetch's life. One of her guests said, "Jean, can we come back next week?" The next week, they brought three more fat friends. The week after that, four additional fat friends joined them.

If this sounds to you like the beginnings of a true-life fairy tale of one woman fighting the odds to attain personal beauty, celebrity, and vast wealth, you're right. Within two months, the weekly meeting had swelled to forty women. A year after the "When are you due?" question, Nidetch was down to 142 pounds. One night, after one of her increasingly popular meetings, a businessman who'd lost 40 pounds thanks to Nidetch suggested she turn her "little project" into what it so clearly deserved to be—a business. She did. Within five years, 297 classes were being held in New York City alone, and there were 25 franchises in 16 states. In 1978, H. J. Heinz, the company that makes the famous ketchup, bought her business for $72 million, making Jean Nidetch the Horatio Alger of weight loss. You've probably heard of it. You may have even heard this near-mythical story before. Jean Nidetch named her company Weight Watchers.

NIDETCH'S SOLUTION to weight loss lay in collective willpower. Weight Watchers wasn't the first diet to push this method. Overeaters Anonymous, which is also based on group support, was founded three years earlier, in 1960.

Group support was just one way people could lose weight. The year after Weight Watchers launched, a high-living photographer put the opposite spin on dieting with *The Drinking Man's Diet: How to Lose Weight with a Minimum of Willpower*, which sold more than two million copies. It was joined that same year by another liquid solution to trimming down: Diet Pepsi. A few years later, a British biochemist introduced the Cambridge Diet, a tough-love, low-calorie regimen designed to promote fat burning and shed pounds fast.

The pace of diets and dieting was starting to pick up in the 1960s. People were getting fatter. According to the Centers for Disease Control, in the early 1960s, just 13.4 percent of adult Americans qualified as obese. A decade later, the percentage had ticked up more than a full point to 14.5 percent. (The increase during this period is even greater when obesity is measured by skin fold rather than the more simple body mass index calculation.) Obesity really got rolling, however, in the '80s, and by the late '90s, more than 30 percent of American adults were obese, more than double the early '60s tally.

All that dieting, in other words, didn't work. Despite Jean Nidetch's life-changing insight, and the true-life miracles behind every weight-loss regime since, we continue, year after year, to gain weight. In Jean Nidetch's day, obesity was a relatively rare condition. Now it's common. Today, obesity is holding at 35 percent, nearly triple what it used to be. By the mid-2000s, the 1961 Jean Nidetch, with a BMI of 33.5—squarely in the midrange of "obese"—would have looked almost normal. Today there is extreme obesity, which hardly existed in the early '60s. Back then, just a tiny slice of Americans met this qualification—0.9 percent. The "pregnant" Nidetch was herself forty-one pounds shy of that mark. Today it's at 6.4 percent.

To put this in perspective, at a sold-out Pirates-Yankees World Series game in 1960, there would have been around six hundred fans in Yankee Stadium of a girth that verged on shocking. Today, there would be close to forty-five hundred, and no one is shocked by it. In the early '60s, well over half of Americans were "slender" and of the nonslender, the vast majority was classified as "overweight"—they needed to lose a few pounds. It is now abnormal to be slender. Today, less than a third of Americans are slender, which is another way of saying more than two-thirds are either overweight or obese. Ninety million Americans—the populations of greater LA, New York, and Chicago multiplied by 2—now eat so much they are at increased risk of asthma, cancer, heart attack or stroke, reduced fertility, giving birth prematurely, high blood pressure, sleep apnea, liver disease, gallbladder disease, diabetes, and arthritis. The obese make less money (particularly obese women), have higher medical expenses and lower self-esteem, and are more likely to suffer from depression. After smoking, obesity is the leading cause of preventable death. And when it comes to morbidity—"a diseased state or symptom"—obesity is surging past smoking, drinking, and poverty.

Obesity is so rampant that it seems contagious. It's an epidemic now, and it's spreading to other countries—the British are gaining, the Chinese are gaining, even the French are gaining—which makes it a pandemic. There are frantic efforts to make it stop. Weight Watchers and Overeaters Anonymous were just early tactics in a long war that would go on to include the Pritikin Principle, the Scarsdale Medical Diet, Slimfast, the Atkins Diet, the South Beach Diet, The Zone, Nutrisystem, Jenny Craig, the Blood Type Diet, the Mediterranean Diet, the Master Cleanse, the DASH diet, the Cabbage Soup Diet, the Paleo Diet, and the Raw Diet. Americans have eaten fat-burning grapefruits, consumed cabbage soup for seven straight days, calculated their daily points target, followed the easy and customizable menu plan, dialed the 1-800 number to speak to a live weight-loss counselor, taken cider vinegar pills, snacked strategically, eliminated high-glycemic vegetables during the fourteen-day induction phase, achieved a 40:30:30 calorie ratio, brought insulin and glucagon into balance, sought scientific guidance from celebrities, abstained from the deadly cultural practice known as cooking, tanned and then bled themselves to more fully mimic the caveman state, asked that the chef please prepare the omelet with no yolks, and attained the fat-burning metabolic nirvana known as ketosis.

It has all been a terrible, amazing failure. The average American man has gained twenty-nine pounds and the average woman twenty-six. Between 1989 and 2012, according to the market report "The U.S. Weight Loss & Diet Control Market," Americans collectively spent more than $1 trillion on weight loss. In that same period of time, obesity grew by more than 50 percent and extreme obesity doubled. The long battle against weight gain hasn't been much of a battle—more like trying to put out a forest fire with a garden hose.

What a strange problem. Despite living in a culture that prizes thinness above even wealth, we keep on eating. It's as though we've created a new "diet-resistant" form of obesity that, like some kind of cancer, perpetuates itself at the expense of our own vitality. Kindergarten children now struggle with their weight. Fully one-third of boys and girls from six to nineteen years of age are overweight or obese.

And obesity is just the most visible manifestation of a deeper malaise. Food has become a life-threatening indulgence. It seems to be disrupting the very way our bodies run—straining our organs, distressing our bowels, and crashing our mood. Adult-onset diabetes had to change its name to type 2 diabetes because so many children are now being diagnosed with what was formerly considered a metabolic disease of grown-ups. Once upon a time, we ate to sustain ourselves. Now food itself is toxic.

What happened?

SUGAR. That's the latest answer, anyway. As I write these words, sugar— or "white death," as some have taken to calling it—is igniting flares of panic and condemnation. A year or two ago, a panic over high-fructose corn syrup came through like a flash flood and then died down to a trickle. Saturated fat, which used to be deadly, is enjoying a renaissance while polyunsaturated fat, which at one time was seen as the antidote to saturated fat, is now under attack. Before fat it was carbs and before carbs it was fat, and if you go back far enough sugar pops up again. For the better part of a century, millions of people, almost all of them with a rudimentary or nonexistent understanding of biochemistry, have been taking part in a richly technical conversation about such phenomena as glycemic load, protein ratios, and serum triglycerides.

Part of the problem is human nature. We are all natural reductionists. We always want to find the single cause of this or that problem, because then it's easy to come up with a silver-bullet solution. That sort of thinking works very well when it comes to car trouble—your alternator is fried, your air filter is clogged, your timing belt is worn out. (If it's all three, it's time for the scrap heap.) But it doesn't work very well with nutrition, which is about a lot of things. The list of essential vitamins, fats, and amino acids includes twenty-four different substances. And that doesn't include minerals, trace minerals, fiber, choline, or the very fuel of life: energy. But even when you add those to the list, along with starch in all its amazing forms and the micro-universe of fats, you still haven't come anywhere close to describing the radiant complexity of the plant and animal matter that goes into our mouths, our stomachs and intestines, and eventually becomes part of our bodies.

That's the other problem. Food is complicated. And when a species that delights in one-word answers faces a problem as complex but crucial as food, the result is not surprising: a decades-long kangaroo court in which we keep putting the latest evil nutrient on trial. The truth is, it would all be so much simpler if it really were just sugar's fault.

But clearly, something—or things—did change.

Here's one thing that definitely did not change: our genes. This is not evolution. There was no cataclysmic event—no meteor, no supervirus that wafted out of some secret government lab—that conferred a reproductive advantage to those inclined to obesity. Similarly, there has been no demographic influx of genetically obese immigrants who fundamentally changed the population. Make no mistake, there are genetic aspects that determine each individual's propensity to obesity—I might be more susceptible to putting on weight than you because of traits I inherited from my parents. But as a group, we all have pretty much the same genes as we had in the 1960s. And that can mean only one thing: Something in the world around us has changed.

When you stop to think about it, the human body faces the same doozy of a problem as the nutritionists. It has complex needs. And it fulfills those needs with a very complex substance: food. How does it do that? How does a body know what it wants?

That, it turns out, is the part we've been messing with—the *want* part. Sugar has something to do with want, and so do high-fructose corn syrup, fat, carbs, and all those other nutrients we've been obsessing over. But the cause of the food problem will not be found in individual nutrients. We keep mistaking the mechanism of obesity for the cause. If we regarded smoking the same flawed way we understand food, we would say cigarettes are deadly because they cause cancer. Cancer is the *how* of tobacco-related mortality. The reason people smoke in the first place—the *why*—is that tobacco is addictive. People smoke because they experience a powerful *desire* to smoke. Jean Nidetch's problem, similarly, was behavioral. It wasn't that her body turned all the food she ate into fat, or that perhaps it was exquisitely efficient in turning refined carbs into fat. That's what bodies do. Her problem was that she ate too much food. She wanted to eat. She could not resist the desire. And when it comes to wanting, food speaks its own special language: flavor.

Flavor, as we will see, is the aspect of the human environment that has changed. The food we eat today still seems like food, but it tastes very different than it used to. For the better part of a century, two complimentary trends have conspired to transform the flavor of what we eat. These two trends were already ascendant when Jean Nidetch was mistaken for pregnant in that Long Island supermarket. And within a year, they would unite in a Dallas suburb with the momentous utterance of a single word: "taco."

This is where our story begins.

———————

IN THE SUMMER of 1962, the vice president of marketing at Frito-Lay took his wife and three kids on a trip to Southern California. It was, on the surface, a family vacation. The five of them piled into Dad's gold Lincoln Continental for the long trip from Dallas to Orange County, stopping along the way at Carlsbad Caverns and the rim of the Grand Canyon.

From the very beginning, however, the trip portended big things about flavor. Before getting hired by Frito-Lay, Arch West had been a Madison Avenue ad man, where he'd headed up the Kraft account and worked on Jell-O puddings. In Corona del Mar, the West family stayed at a house belonging to Lawrence Frank, the inventor of Lawry's seasoned salt. And one afternoon, after the family had just dined at a restaurant called the Five Crowns—West liked the prime rib and fancy creamed spinach—a stranger walked up and complimented his daughter's golden blond hair. The man asked the Wests if they'd ever eaten at his restaurant, but they'd never heard of it, even though in just two years the 500th location would open in Toledo, Ohio. The man was Ray Kroc and his restaurant was McDonald's.

The most important meal of that trip, however, didn't take place at the Five Crowns, or at the restaurant that would go on to become the world's largest chain of hamburger fast-food restaurants. It was served at a little Mexican "shack" West spotted by the side of the highway somewhere between L.A. and San Diego, where he pulled over and ordered a small container of tortilla chips.

It was likely the crunch that got him. Besides shape, crunch is the only aspect in which tortilla chips are meaningfully different from a snack West was already in charge of marketing, Fritos. Both are fried pieces of cornmeal. Tortilla chips, however, are baked first, which makes them crunchier. Arch West was struck by an idea: Tortilla chips just might be Frito-Lay's next big thing.

Back at company headquarters in Dallas, West presented his great new idea to his fellow executives. The response was something like the sound a vacuum cleaner makes when it's unplugged. Why would Americans want Mexican "tortilla" chips, his colleagues wondered, when they already had perfectly good corn chips? They weren't even interested in trying one. Their instructions were clear: Do not pursue tortilla chips.

West knew better. He was so confident about the future of tortilla chips that he secretly funneled discretionary funds to an off-site facility to develop the tortilla chip concept. He pitched his idea again. This time, though, he handed out samples. He had a plan: They weren't tortilla chips anymore. Now they had their very own Mexican name, one that meant little pieces of gold: "Doritos." West got the green light.

The rest, however, is not history. The Doritos people all over the world know and love, and gobble four at a time, almost never happened. The Doritos Arch West used to seduce his fellow executives, and that would hit store shelves in 1964, were exactly like the ones West tasted back in California, just salted tortilla chips—"toasted corn taste" is how they were billed on the packaging. They sold decently in the Southwest, where people knew that the pointy tip was well suited to scooping up globs of dip. (The early packaging even featured an illustration of a hand dipping a Dorito in dip.) But the rest of the country didn't know what to make of them. Doritos sounded Mexican, but they didn't taste Mexican. This was a problem.

Archibald C. West once again found himself facing his fellow executives over his catchy new snack—a snack he wasn't even supposed to develop—that wasn't catching on. West didn't give up. Instead, he uttered the word that changed everything. Make Doritos, he said, taste like a "taco."

The Frito-Lay executives sneered. As his son Jack West recounts, they chided the fancy New York pitchman for "not knowing the difference between a 'thing' and a 'flavor.'" But West was one step ahead of them. Perhaps because of his friendship with the Lawry's seasoning mogul,

West already knew that the line between things and flavors could be blurred, that technology existed that could impose the flavor of a taco on a fried triangle of corn. "Of course, you and I know that," he fired back, "but the rest of the country north of here sure doesn't. And that's our market."

And what a market. The Northeast, the Northwest, the South, the Southwest—everyone loved taco-flavored Doritos. They loved them so much that four years later, Frito-Lay blurred the line between thing and flavor once again, this time with Doritos that tasted like nacho cheese. In 1986, Cool Ranch—a tortilla chip flavored like salad dressing—was born. By 2010, the chip beloved by everyone from toddlers and teenagers to stoners and the infirm was earning Frito-Lay $5 billion a year. There are, at present, fourteen flavors of Doritos in the United States, including Salsa Verde and Spicy Sweet Chili. Every day around the world, fingers numbering in the tens of millions become coated in sticky orange seasoning. Every second, untold trillions of neurons are fired by that irresistible combination of salt, fat, and flavor while the people attached to those fingers experience the irresistible desire to put their hand back in the bag for more. "One good crunch," as the 1968 package copy trumpeted, "leads to another . . . and another."

"Taco."

A THING, of course, is different from a flavor. Different things have their own different flavors. Oranges taste like oranges. Bananas taste like bananas. Tacos taste like tacos, and corn chips taste like corn chips. That, at least, is how the world worked back when there were still families who'd never heard of McDonald's.

Years before West arrived at Frito-Lay, the company launched "barbecue"-flavored potato chips, a breakthrough that made it possible to give fried slices of potato some of the same smoky, sweet notes as meat cooked slowly over hardwood. People who ate barbecue chips liked washing them down with soft drinks that tasted like oranges, grapes, or lemons, even though these foods contained none of these "things." By the early 1960s, however, flavor technology had taken a great leap forward. The science was now so good it was possible not just to blur these lines but to utterly distort them. And that's what West did. He gave a simple fried piece of corn the tang and savory depth of a Mexican meal.

"Things," meanwhile, were also changing. Fruits, grains, meat, and vegetables were themselves losing flavor. The corn Frito-Lay used to make Fritos in the 1960s looked just like the corn Elmer Doolin used when he founded the Frito Company back in 1932. But it didn't taste the same, because by 1967 an American corn farm was growing nearly three times as much corn as it had thirty years earlier. There was more corn, but it tasted weaker, like a lesser version of itself. Corn was getting bland. So were potatoes. The same year Elmer Doolin started making Fritos, Herman W. Lay got into the potato chip business. Back then, a typical American potato farmer produced about sixty-three sacks of potatoes for every acre. By the mid-1960s, it was up to two hundred sacks. And just like corn, the potatoes in those sacks didn't taste as "potatoey."

That problem could be solved. The gathering void of blandness was filled by industry. Using the most sophisticated analytical technology of the era, scientists isolated the mysterious chemicals that humans experience as flavor, and the companies they worked for began manufacturing them and selling them to food companies, which added them to their products. You can see those chemicals right there on a 1968 package of taco Doritos represented by a single, exceptionally vague word: "Flavorings."

West's genius was one of vision. He stood firmly astride two waves—food getting blander and flavoring getting better—and married them. He showed how extraordinarily potent flavor technology had become. Taco Doritos tasted better than salted Doritos. And, unlike actual tacos, they didn't spoil, they were never overdone, they always tasted the same, they didn't need to be cooked, and they were cheap. A 1¾-ounce bag of original Taco Doritos sold for 15¢.

The Dorito didn't just predict the future of tortilla chips. It didn't just predict the future of snack food, either. It predicted the future of all food. Nothing tastes like what it is anymore. Everything tastes like what we want it to taste like. As food gets blander, we crank out zestiness by the hundreds of tons to make up for it. Most people recognize this as junk food. But it's happening to food served at restaurants and the food people buy at the supermarket and cook, from scratch, at home. It's happening to blueberries, chicken breast, broccoli, and lettuce, even fennel. Everything is getting blander and simultaneously more seasoned. Everything is becoming like a Dorito.

The birth of Doritos was a watershed moment. Flavor wasn't up to Mother Nature anymore. Now it was in the hands of the folks in marketing.

FOR ALL its technicality, the food conversation has been strangely silent on the topic of flavor. Back in Arch West's day, no one thought ingredients like torula yeast, flavorings, or MSG were particularly dangerous, and that thinking hasn't changed much today. They're noncaloric, for one thing. You could never get fat on a diet of these chemicals. They don't cause cancer or debilitating brain disease, either. (Not directly, at least.) Is there even any point in scrutinizing pleasure? Hedonism, as any puritan can tell you, never leads to virtue. If we could all set pleasure aside and eat what's good for us, our problems would all go away. (Good luck with that.)

Let's not be too quick to lay all the blame on a 1960s snack food executive. The man who invented Doritos was a World War II veteran and churchgoing family man who was raised in a Masonic home and once got injured while volunteering for disaster relief when his car was hit by a tanker truck. Arch West, furthermore, understood something the field of nutrition and the $60 billion weight-loss industry have only recently showed the faintest glimmer of grasping: Flavor matters. Eating isn't a rational act of nutrient acquisition. Eating has as much to do with nutrients as sex does with procreation—we do it because it feels good to do it. We might pretend we're interested in vitamins, fish oil, and ketosis, but it's flavor we're after. We think in flavor, we dream about flavor, and we get up out of our chair when the bases are loaded in the bottom of the ninth to get it. We eat for one reason: because we love the way food tastes. Flavor is the original craving.

This is not because we are lazy or weak. It's by design. If you think of the human genome as an instruction manual with each bodily system having its own chapter, you will discover something quite unexpected. The thickest chapter is the one on flavor. Our ability to sense and enjoy food is no accident. Not only are we expert flavor sensors, but the flavors we sense have a firm grip on our minds. They drive our behaviors and control our moods. If music is emotion expressed in the medium of sound, flavor is emotion expressed in the medium of food.

We are, if you like, playing with our own minds. And our game has gotten a lot better since Arch West's taco moment. Taco Doritos listed eleven ingredients. The much more recent Jacked Ranch Dipped Hot Wings Doritos—a tortilla chip that tastes like chicken wings dipped in hot sauce and then dipped in salad dressing—lists thirty-four.

So imagine for a moment an alternate world in which everyone is wearing flavor goggles. When they bite into foods that taste like tacos, cherries, grapes, or oranges, their brains think they are actually eating tacos, cherries, grapes, or oranges. But what they are actually tasting are flavor chemicals.

That's the world you live in. You may not think so. You may believe you possess the kind of sophisticated palate that can easily spot the difference between a real taco and a taco-flavored tortilla chip, or between a real grape and a grape-flavored beverage. Your flavor-sensing system, however, is being fooled. And the proof is in the fact that you—

we, all of us—*like* these flavors. We liked taco-flavored Doritos more than plain ones, even though we know they're not really tacos. We like Coke, 7Up, and ginger ale more than plain old sugar water. And we like the flavor chemicals we didn't even know are being added to apparently wholesome foods, like raw beef, butter, soy milk, yogurt, and tea. The deception is so elegant as to be invisible. We are all wearing flavor goggles.

There are consequences. Bland, synthetically flavored food is not the same as naturally flavorful food. On the most basic level, when real foods like tomatoes, strawberries, and chicken taste bland, we make them palatable the only way we know how, by pouring ranch dressing over watery tomatoes, ladling dollops of whipped cream over strawberries, and blitzing chicken in flavoring and then dunking it in the deep fryer. We do what Arch West did to plain old tortilla chips to get people to eat more of them.

It gets worse. Nature endowed us with our most sophisticated bodily system because it performed the body's most essential task: getting important nutrients. By manipulating our richest and most direct source of pleasure, we have warped our relationship with the fuel our bodies require, food. Evolution may have given us an amazingly complex flavor-sensing apparatus, but it wasn't made for a world of cheap calories and egregious flavor lies. We have taken a system designed to bring our bodies to a state of nutritional completion and turned it against us.

The Dorito Effect, very simply, is what happens when food gets blander and flavor technology gets better. This book is about how and why that took place. It's also about the consequences, which include obesity and metabolic disturbance along with a cultural love-hate obsession with food. This book argues that we need to begin understanding food through the same lens by which it is experienced: how it tastes. The food crisis we're spending so much time and money on might be better thought of as a large-scale flavor disorder. Our problem isn't calories and what our bodies do with them. Our problem is that we want to eat the wrong food. The longer we ignore flavor, the longer we are bound to be victims of it.

This book is also about the solution. The Dorito Effect can be reversed. That's already happening on small farms and in pioneering science labs. Not only can we imagine a world where the food tastes better and people eat less of it—we can also visit it. I have visited it, and the food, as you will see, is superb. One day, we may look back on this obesity epidemic as a curious aberration in history when advances in analytic and synthetic chemistry outpaced our knowledge of psychology and nutrition.

If you like the food in the world you're in just fine the way it is—the way it tastes, the way it makes you feel, and the effect it has on your body—this book isn't for you. Get your money back before the spine creases and spend it on bland, synthetically flavored food. I'm confident you'll enjoy eating it; I'm less confident you'll enjoy the consequences. If you want to discover the true nature of our relationship with food and how we've manipulated the ornate chemical system that sparks cravings and touches every cell our bodies, turn the page.

TWO

Big Bland

To THOSE who doubt the blandness of modern chicken, consider the following story:

In the town of McPherson, Kansas, there is a butcher shop called Krehbiels Meats, where, not long ago, an elderly woman bought a chicken that moved her to tears. She spotted it in one of the display cases, a brand she hadn't seen before called Good Shepherd Poultry Ranch. The chicken beneath the shrink-wrap packaging had longer legs, a smaller breast, and yellower skin than regular chickens, and on the back appeared two words the woman, who was in her seventies, would not have seen in a very long time: "barred rock." This chicken was a throwback, a variety nearly vanished since the 1950s and still raised the old-fashioned way, outdoors, where it ate blades of grass, leaves, seeds, bugs, and whatever else it could put its beak on. The woman purchased one and took it back home, an hour south to Wichita.

She had every reason not to be excited. During the course of her forty-eight years of marriage, chicken had only ever brought disappointment. The problem was chicken and dumplings. It was one of her husband's favorite dishes, but every time she made it his verdict was always the same: "Not as good as my mother's." She tried different recipes.

She tried different ingredients. But as the years turned into decades, his judgment never wavered. After close to a half century of marriage, she was married to a man who still missed his mother's chicken. And that, it seemed, is how it was fated to be.

For whatever reason, the woman decided to give chicken and dumplings one more try with the barred rock chicken from Krehbiels. This time, her husband was astounded. This time, he swallowed his dumplings and delivered the news she'd been waiting almost fifty years to hear: "This is my mother's chicken and dumplings."

It was at this point in the story that the woman began weeping. She was on the phone with the farmer who raised the chicken, an heirloom poultry buff named Frank Reese, whom she'd never met or spoken to before. As the tears flowed, she said, "I just wanted to call and thank you so much," and she shared memories of eating chicken on a farm as a little girl. When they hung up, Reese went back outside to the thing he loves most of all: looking after his barred rock chickens.

The woman's mistake, it turned out, had nothing to do with the recipes she'd been using. It had nothing to do with her cookware, her oven, the thickness of her dumplings, her gravy, the amount of salt and spices she used, the hardness of her tap water, or any of the usual variables a home cook might place under the beam of doubt. All that time, the problem was the flavor of the chicken itself.

It's a complaint we hear often from the blue-hair set. Nothing tastes the way it used to. We tend to dismiss it as the rose-tinted memory of times past or the result of failing taste buds. But the blue-hair set is on to something. Food has changed. The change has been documented scientifically. And it is a story best told by chicken, which has become not only our number one source of animal protein but is simultaneously the blandest and most flavored—the most Dorito-like—meat.

———————

CHICKEN'S DESCENT into blandness began precisely sixty-one years earlier, in March 1948, when fifty world-changing chicks pecked out of their shells at a hatchery in Easton, Maryland. The eggs had come all the way from Marysville, California, where a month earlier, at an outfit called Vantress Hatchery, a fine plump California Cornish rooster mounted some fine plump red New Hampshire hens. They arrived in Easton along with 31,630 eggs from 25 different states and were there as part of a grand event that would determine the way soups, broths, and braises of chicken would taste for decades to come.

The Chicken of Tomorrow contest was conceived by Howard "Doc" Pierce, who was national poultry research director with A&P Food Stores, one of the largest grocery chains of the time. For anyone in the chicken business, the late 1940s was the best of times, but also the worst of times. World War II, which had just ended, had been good to chicken farmers. As red meat was rationed, Americans almost doubled the amount of chicken they ate. But with the war over, Pierce worried that the spike in chicken consumption would come to a crashing end as Americans returned to their red-meat-loving ways. Pierce wanted to stop that from happening.

Chicken of the 1940s was nothing like it is today. It was expensive by modern standards, and since chickens were often the by-product of the egg industry, they came in a range of sizes. There were broiler chickens, which were young and tiny—some weighed in at just a pound and a half—and so tender you could cook them under a scorching-hot broiler. Next came fryers, which were a bit bigger and less tender, but still small. After fryers came roasters, and last came "fowl"—old hens that were so tough they could be used only in soups and stews. If a quick and easy Tuesday night dinner was what you had in mind, you needed a broiler or a fryer. You might even need two. And it was going to cost you.

What this country really needs, Pierce thought, is a steady supply of tender, large-breasted chickens. So A&P put up $10,000 in prize money and sent wax models of perfect-looking chickens around the country. Whoever could raise the flock of chickens that grew the fastest and looked most like the wax model stood to make quite a bit of money.

In 1946 and 1947, regional Chicken of Tomorrow contests were held. The cream of that group was invited to compete in the national event in 1948, which is how 31,680 eggs from 25 different states found their way to a hatchery in Maryland. Once hatched, the chicks were raised in identical pens and fed a secret diet that contained a minimum of 20 percent protein, 3.5 percent fat, and 7 percent fiber. After twelve weeks and two days, the chickens crossed the metaphorical finish line—they were slaughtered. Then came judgment. Under bright and unflattering light, their plucked bodies were scored for things like uniformity of size; quality of skin; the length, depth, and width of the breast; along with performance traits such as hatchability, feed efficiency, and average weight. Those Vantress chickens were big, averaging 3.75 pounds, and scored 3.17 for feed efficiency, which means it took just over three pounds of feed for every pound of chicken.

This is what agricultural types call "improvement." And it had been going on at a slowish pace for a couple of decades already. Back in 1923, it took sixteen weeks to get a chicken to a relatively puny live "broiler" weight of 2.2 pounds, with a feed efficiency of 4.7. By 1933, that same broiler had gained half a pound, and took two fewer weeks to do so. By 1943, broilers were averaging 3 pounds at twelve weeks. These Vantress chickens, however, were something to behold. Not only were they roughly a full pound heavier than their peers, they somehow managed to get that big on less food. These were miracle chickens.

How did these miracle chickens taste? No one knows. The judges didn't measure flavor. The point of the contest, after all, was to create a chicken that looked like a wax model.

THERE WAS a second great legacy of the Chicken of Tomorrow contest. Like the Vantress Hatchery chickens, it was a precocious, high-performing biped that would go on to change the world. This one, however, didn't have feathers.

Paul Siegel, who would grow up to become one of the most important and prolific researchers in the history of poultry science, was fifteen years old in 1948. He lived on a thirty-two-acre farm near Vernon, Connecticut, where, from an early age, he displayed the unmistakable signs of poultry genius. When Siegel won Connecticut Poultry Boy of the Year, he got caught playing hooky. His school, phoning to share the good news, thought Siegel was home sick only to discover he was out cleaning the chicken house. The day Siegel learned there was a youth division of the Chicken of Tomorrow contest, he went out to his tiny broiler house and put some New Hampshire roosters with some White Plymouth Rock hens. When the chicks hatched, he kept the feed trough full and made sure not to let them go outside, because any forays they took outdoors were a waste of valuable energy they could be putting into growth. After twelve weeks of fervent poultry husbandry, the county agent pulled up to the family farm to assess the young man's chickens—Siegel can still remember him marking their weights on a score sheet he hung on the back of a wooden door. Some time later, he received the good news. Siegel took the prize for the junior Chicken of Tomorrow contest for the state of Connecticut. Nearly seventy years later, the plaque still hangs in his den. But Siegel wasn't surprised. Being a star poultry boy is like being a star athlete. "I knew I was pretty good," he says.

Siegel's singular talent was exploiting the very principle that had been demonstrated at the Chicken of Tomorrow contest, and that would go on to doom the flavor of chicken and dumplings for decades to come: Chickens can be changed through breeding. Growth rate and plumpness weren't fixed physical laws, like the speed of light or the electrical charge of a proton. By choosing which rooster got to mate with which hens, you could change chickens' genes. You could make chicks that were different—much different—from their parents.

Those genes kept changing. By 1951, the Chicken of Tomorrow winners got fat two weeks earlier than they did in 1948, and by 1955 the winning chickens of 1951 were just average. By 1973, it was down to eight and a half weeks.

Everything, in other words, went according to plan. As World War II concluded, chicken consumption did indeed decline, just as Doc Pierce had feared. But then chickens got cheaper and plumper, and the eating of said chicken rebounded, rising back up to wartime levels in the early '50s and then exceeding them by the middle of that decade. By 1967, Americans were eating twice as much chicken as they had in 1948, and by 2006, chicken had become so cheap and so abundant that Americans were eating nearly five times as much as they had in 1948.

The former Connecticut Poultry Boy of the Year is partly to thank. Six years after winning Connecticut's junior prize, Siegel received a master's degree in poultry genetics. After he got his PhD, he joined the faculty at Virginia Tech, where, after more than fifty years, he's still at it. Over his incredible career, he has personally incubated and hatched around 200,000 baby chicks and published more than 500 scientific papers (a chick-to-publication ratio of 400:1). He is an inductee of not only the American Poultry Industry Hall of Fame but the International Poultry Hall of Fame as well. Pretty good for a guy who played hooky.

The Chicken of Tomorrow contest never really ended. Siegel dedicated his academic career to exploring how chickens could be changed and improved through breeding. He found that there is always a tradeoff. For example, there is a genetic tradeoff between body weight and egg laying. That's why the chicken industry is now split into two distinct halves, meat chickens and egg chickens. Today's meat chickens are pathetic layers compared to today's egg chickens, and the reason is that they put all their energy into creating flesh. Today's egg chickens, by comparison, are so scrawny that no one bothers raising the males into fryers. Instead, as soon as they can be "sexed," the boys are separated from the girls and exterminated—their necks are snapped, or they're gassed, or fed, live, into a high-speed meat grinder.

But Siegel's biggest influence on America, not to mention the rest of the planet, was his students, the small army of poultry geneticists— professional chicken breeders, basically—who went out into the world with a single goal: to improve chickens, to make them ever plumper, younger, more efficient, and faster. In the early years, Siegel's students were hired by the quaint little hatcheries that still dotted the country-side. As chicken consumption spiked and the chicken business became serious business, his graduates joined huge multinational companies whose specialty is poultry genetics. Today there are three global giants: Hubbard, Cobb, and Aviagen. Almost no one has heard of these compa-nies. But everyone has eaten their chicken.

The result has been formerly unimaginable "improvement." The broiler of today looks plumper. The broiler of today grows in less than half the time, about thirty-five days, as the world's fastest chickens did in 1948. Somehow, the broiler of today manages to weigh a pound and a half more and, even more incredibly, gets to that weight on a third less feed. Its legs are so short and plump that chickens, which were once agile goose-steppers, now waddle, and its breasts are so broad and thick that modern chickens don't quite stand up straight. Today's raw chicken is the porn star of the meat world: sensationally curvy and expertly de-nuded. When Siegel thinks back to those perfect wax chickens in 1948, he laughs. "They could never dream what the broiler of today would look like," he says.

The dream of Doc Pierce, in other words, has been gloriously real-ized. Chicken is number one. The country that formerly preferred beef now eats 26 billion pounds of chicken every year. Chicken is cheap. The bird that was selling for 60¢ a pound in 1948 was down to 39¢ in 1968. In 1948, a five-pound chicken cost $3—which might sound cheap, but in 2014 dollars it works out to $30 for a single bird. In 2014, a supermarket chicken will run you $7. Chicken today costs less than a quarter of what it did during the Chicken of Tomorrow contest.

They are all broilers now. Words like "fryer" and "roaster" still appear in cookbooks, but they don't exist anymore. We eat gigantic babies. As a paper in the journal *Poultry Science* puts it, if humans grew as fast as broilers, "a 3 kg (6.6 lb) newborn baby would weigh 300 kg (660 lb) after 2 months."

Paul Siegel likes eating chickens as much as he likes studying them. But he will admit that for all their improvement, they might not taste the same as they used to. The best chicken he's ever eaten wasn't a chicken of tomorrow. It was a chicken of yesterday: his mother's chicken fricassee, which he misses in much the same way the husband in Wichita missed his mother's chicken and dumplings. The tastiest chicken he's eaten since wasn't a commercial broiler. It wasn't even American. It was an eighty-four-day-old slowpoke sold under the brand Label Rouge. And as its exotic name suggests, Siegel ate it in France.

ONE OF the first signs there was a flavor problem with chicken appeared in a cookbook published thirteen years after the Chicken of Tomorrow contest. "Modern poultry raising," it stated, "has done wonders in making it possible to grow a fine-looking chicken in record time and to sell it at a most reasonable price, but rarely does anyone in the country discuss flavor. If you are interested in price alone," it continued, "you will often end up with something that tastes like the stuffing inside a teddy bear and needs strong dousings of herbs, wines, and spices to make it at all palatable." It's pretty clear what was happening. Chicken was getting blander. It needed flavoring. Chicken was turning into something that didn't even exist yet: a Dorito.

The book was *Mastering the Art of French Cooking*, which would go on to become one of the landmark cookbooks of the last century, and one of whose authors was the soon-to-be-famous Julia Child. No one, however, seems to have taken the book's chicken warning to heart. By 1997, the chicken situation had gotten worse. "With the emergence of the modern poultry farm after World War II," an updated *Joy of Cooking* stated, "both the quality and safety of our poultry were compromised. Chicken has suffered the most." A year later, in *How to Cook Everything*, *New York Times* columnist Mark Bittman all but wrote the epitaph for chicken flavor. Chicken "leaves you in the same position you're in when you're cooking pasta: You must add flavor."

In 1961, Julia Child and her coauthors stated that chicken "should be so good in itself that it is an absolute delight to eat as a perfectly plain, buttery roast, sauté or grill." Thirty-seven years later it was "downright bland" and "essentially a blank slate." It wasn't just the occasional disappointing chicken that tasted like the stuffing inside a teddy bear. Now all chicken tasted that way.

To understand exactly what happened to chicken, it helps to acquaint oneself with a dispute that raged several years ago in *Organic Gardening* magazine, which, on the surface, had nothing to do with flavor. In the December 1999 issue, senior editor Cheryl Long wrote an open letter of the charged and inflammatory variety to the secretary of agriculture. Food, she claimed, seemed to be getting less healthy. Long backed up her claim with a recent study published in the *British Food Journal* that compared fruits and vegetables grown in the 1930s and the 1980s. Wholesome things like rhubarb, bananas, and parsnips, the study found, contained fewer of the essential micronutrients necessary to human life than they used to. Calcium, Long pointed out, was down by 19 percent, iron down by 22 percent, and potassium by 14 percent. "Mr. Secretary," Long wrote, "what's going on here?"

In a letter back to Long, the director of the department's Research Service admitted that, yes, it did look like the nutritional content of fruits and vegetables was going down, but not to worry because there could be all sorts of reasons, including imprecise measurements back in the 1930s, and that it might not matter anyway. *Organic Garden-*

ing returned fire with an editorial titled "As Food Quality Drops, the USDA Just Shrugs." At which point, the USDA shrugged and the world returned to its nutrient-depleted vegetables and teddy-bear-stuffing chicken.

Everyone, that is, except a biochemist by the name of Donald Davis who worked at the Biochemical Institute, the University of Texas lab with the distinction of having discovered more vitamins than any other lab in the world. Davis read the British study and was duly alarmed, but he noticed a problem: It didn't account for moisture. Some modern fruits and vegetables, he noted, contained more water than the old ones. The problem might be nothing more than the fact that modern produce was plumper and juicier than heirloom produce. So Davis and two colleagues set out to compare the nutritional properties of thirty-nine vegetables, three melons, and strawberries from 1950 with the very same readings performed in 1999, only this time, they adjusted the results to reflect differences in moisture. They also used a more sophisticated statistical methodology (medians instead of geometric means, but never mind) and generally bent over backward and sideways to make sure there wasn't the slightest hint of bias against the modern produce.

Once again, there were differences. Startling ones: 1950s kale had twice as much riboflavin as modern kale, 1950s cauliflower had twice as much thiamin, and 1950s asparagus had almost three times as much ascorbic acid (vitamin C). The trend wasn't universal—1999 green onions, for example, had more riboflavin than 1950s green onions (but way less calcium), to cite one among a few counterexamples. But the overall trend was pretty clear: ascorbic acid down by 15 percent; vitamin A down by nearly 20 percent. On it went. It was as though modern produce had been nutritionally dumbed down.

Davis and his colleagues put these findings in a paper, which they submitted to the well-respected *Journal of the American College of Nutrition*. As is always the case, the paper was sent to esteemed anonymous scientists around the country for independent review. The initial feedback was good, but one reviewer made a comment that caught Davis's attention. He mentioned something called the "dilution effect." In all his years of nutritional research, Davis had never come across this intriguing term. He began researching it, and what he discovered was indeed intriguing. Scientists had been aware of it since as far back as the 1940s.

What they had noticed more than half a century earlier is that when crops are fertilized and irrigated, they contain lower concentrations of major minerals and trace minerals. But intensive farming wasn't the whole story. Davis also found evidence that there was genetic dilution taking place. Part of the reason things like broccoli, wheat, and corn were losing nutrients was that broccoli, wheat, and corn had changed due to careful breeding. Just like chickens, they'd been selected to grow faster and bigger, and that was diluting the nutrients. It was as though everything farmers had been doing for the last fifty years—breeding, fertilizing, spraying—was ganging up on nutrition.

This is alarming. For one thing, the story of the last fifty to one hundred years of agriculture is the story of massive, world-changing leaps in yield. The explosion in productivity has been so miraculous there's even a term for it: the green revolution. It is thought to have saved more than a billion people from starving to death. But Donald Davis noticed a dark side to the revolution that no one talks about. Although the gains in quantity have been huge, there has been a corresponding loss in quality.

What, you might wonder, are plants replacing all those nutrients with? If we're harvesting millions of pounds of broccoli and that broccoli has less calcium and magnesium in it, what's taking their place? Each plant tells its own story, Davis says, but generally speaking we're getting the following: more water and more carbohydrates.

NONE OF the tiny nutrients Donald Davis tested for will help you throw a better dinner party. Riboflavin may play a crucial role in the decarboxylation of pyruvate (don't ask), but it's flavorless. So are thia-min and niacin. In fact, of all the nutrients Davis studied, the only one that tastes of anything is ascorbic acid, which tastes sour.

This is one of the main reasons that for so long no one has seen much of a connection between flavor and nutrition—essential, life-sustaining nutrients like vitamins don't taste like anything. But the nutritional dilu-tion Donald Davis measured is nonetheless proof of flavor dilution. It is the key to understanding not only how fruits and vegetables got bland but also why Julia Child came to compare modern chickens with teddy-bear stuffing. There is one obvious point of connection: moisture, the variable Davis worked so hard to correct for. If modern peppers, cabbage, and strawberries contain more water, then they're going to taste watery.

Flavor dilution, however, goes much deeper than water, and we know this because in 1989, a molecular biologist working at Monsanto named Harry Klee set out to make a plant that had suffered particularly badly at the hands of flavor dilution—the tomato—taste better. He failed. But failure brought Klee face-to-face with a problem plaguing all of mod-ern agriculture, one that looked very much like the one Donald Davis found. Water was part of that problem, but not all of it.

When Klee set out to fix bland tomatoes, he and most other tomato scientists thought they knew what the issue was: Tomatoes had no fla-vor because they were picked green. Since tomatoes have to make such a long journey from fields in Florida, California, and Mexico to super-market shelves as far away as Michigan, Alaska, and Maine, farmers picked them when they were still green as a frog. Sometimes they rip-ened on the truck, and very often they were stored in warehouses and later fogged with ethylene gas, which advances the ripening process. Hence, the problem. Because a tomato that ripens in a truck or a ware-house, as every grandmother can tell you, just isn't the same as a tomato that ripens on the vine. As soon as it has been severed from its meta-bolic energy source—leaves—it loses most of the capacity to turn itself into something delicious.

Klee had an idea. Why not create a tomato that ripened more slowly? That way, it could cling to its life-giving vine long enough to become almost ripe by the time it was picked, and hit full ripeness on the truck. It might not be quite as tasty as a vine-ripened tomato, but it was sure to be much better than a tomato that was picked green.

This was not a new idea. For years, scientists had thought that slower ripening could fix the tomato. What was new, however, was the technology that could make it happen: genetic engineering. In 1989, Klee and his team inserted a gene from a bacterium found in soil into the genome of an ordinary plum tomato, and it did exactly what they'd hoped it would do. Ripening began taking place slow motion. Instead of taking one week, it took three weeks.

And so, on a sunny morning in June 1990, Harry Klee walked out into a Monsanto test plot in Bonita Springs, Florida, where, three months earlier, rows of tomato seedlings genetically engineered to ripen slowly had been tucked into the dark earth. The plants were now three feet tall, each branch heavy with pink orbs like some kind of subtropical Christmas tree. The fruit had just started ripening. Klee next saw the tomatoes inside a climate-controlled storage facility where they were stored on shelves in cardboard boxes. The effects of that soil bacterium's gene on shelf life were "spectacular"—far better than Klee had dreamed possible. These transgenic tomatoes stayed plump and red for months. Klee reached into a box, pulled out a perfect specimen and, using the long blade of his Swiss Army knife, sliced off a disk and popped it in. His mouth was bathed in sugars and acids and pinged by flavor compounds.

But not enough sugars, acids, and flavor compounds. That genetically engineered, slow-ripening tomato—a tomato that, Klee estimates, cost Monsanto around $10 million to develop—was better than a standard supermarket tomato, which Klee considers "tasteless." It was not, however, the kind of tomato that made him pause, midchew, to reflect on the positive and uplifting event taking place in his mouth. After all that effort, it was just somewhat better than a bad tomato. Progress?

Perhaps. But even with $10 million of help, the tomato was, as Klee puts it, "nowhere close to where we wanted to be."

Whatever was wrong with tomatoes, Klee realized, it went way beyond picking them green. So in 1995, Klee left Monsanto and joined the Horticultural Sciences Department at the University of Florida in Gainesville to immerse himself in tomato flavor the way Paul Siegel had done with chicken breeding. After nearly twenty years and three million tomatoes' worth of flavor research, Klee has scaled back his original certainties. Now the one thing he's certain of is this: Modern tomatoes are very, very bland.

Picking them green is one among numerous insults. And you can't fix it by slowing ripening because when the genes that regulate ripening are turned down, Klee learned, ripening never fully takes place. Delayed ripening is just another way of saying impaired ripening.

Color is another problem. In the early part of the last century, tomato growers seized upon a mutation that made tomatoes uniformly red, which makes them appear luscious and all the more ready to eat. Before this, tomatoes bore patches and stripes of green. They tasted all the better for it, because those stretches of chlorophyll produced energy that powered the flavor-making process. But as with the wax model chickens, the focus was on how they looked, and by the 1950s tomatoes were a factory-finish red.

And then, of course, there is yield. A hundred years ago, a typical tomato plant was twelve feet tall and carried four or five ripe tomatoes at any one time, with a few green babies still weeks away. Around the time of the Chicken of Tomorrow contest, tomato breeders set about cranking up tomato output. A tomato plant now tops out at six feet and carries as many as ten ripe tomatoes at once. That's too many. According to Klee, the plant is "source limited." It doesn't have enough leaves to power all that fruit, so it undergoes the plant equivalent of a brownout. Like a frantic parent, the plant fills its fruit with the only thing it can: water. And the tomatoes taste like what they're filled with.

But even if you fixed the color situation, balanced the leaves-to-fruit ratio, and got rid of all that water, you still wouldn't fix the modern tomato's flavor problem. And the reason is that tomatoes, on a genetic level, have forgotten how to taste good. As breeders selected moneymaking traits like yield, disease resistance, and a thick skin for easier transportation, they ignored the genes that determine good flavor. There are a lot of those genes, and with each generation, some aspect of flavor can be lost. Over uncountable generations, the loss is substantial. And when the flavor genes are gone, there's only one thing that can make a tomato taste good: a bottle of ranch dressing.

CHICKENS, of course, are not tomatoes. They have feathers, not leaves, and broilers are anything but "source limited." They get to stand there at the feeder and gorge. And boy, do they ever gorge. Modern broilers have been intentionally bred to be voracious. These birds don't even want to go outside to eat grass and bugs, because the genetic tradeoff for fast growth—genetics, remember, is all about tradeoffs—is laziness. But there is still a dilution problem. And, just as with bland tomatoes, a chicken's blandness is intimately related to nutrition.

For the first 99.9925 percent of their domesticated careers, chickens ate all sorts of stuff: blades of grass, leaves, seeds, bugs, mice, frogs, meat scraps, dead rabbits, even snakes. Their human keepers carried the barest understanding of a chicken's dietary needs. The Romans thought visiting a dung heap was the thing to do. (Not ridiculous at all: There are a lot of bugs in dung heaps.) By the turn of the last century, the picture wasn't a whole lot clearer. Poultrymen and poultrywomen knew you could fatten chickens by feeding them things like ground-up corn. But they also knew that if all you gave chickens was corn, they'd get sick. So they sent chickens out to forage, and their beaks would find the foodstuffs that kept them healthy. In the winter, chickens would get milk, cabbage, green onions, bran, and table scraps. Without green treats and outdoor foraging, chickens got sick and died. No one knew why.

Around the turn of the last century, a Dutch physician named Christiaan Eijkman observed that when his chickens were fed white rice and only white rice, they became afflicted with beriberi—they had difficulty walking, they would vomit, and eventually they became paralyzed and died. If the sick chickens were fed brown rice, however, they recovered. Eijkman postulated that there must be something about brown rice, some hidden essence crucial to maintaining health, that was not to be found in white rice. A few years later, a Polish biochemist by the superb name of Casimir Funk took the stuff that makes brown rice brown—rice bran—and treated it with alcohol and phosphotungstic acid and was left with a tiny amount of an almost magical substance that could cure a pigeon just hours away from death by beriberi. Funk called this revolutionary substance a "vitamine." (It was, in fact, vitamin B_1, properly known today as thiamin, and which is down by half in cauliflower and collards.)

The study of nutrition would never be the same. Thanks to vitamins, deadly diseases like rickets, scurvy, beriberi, and pellagra would become not only treatable but preventable. Eijkman was awarded a Nobel Prize. (Funk got bupkis.) But what Christiaan Eijkman almost surely did not realize is that thanks to him, the birds he was studying would, before the century was over, taste like teddy-bear stuffing.

As poultry scientists ticked off the list of vitamins, minerals, amino acids, and the other microscopic substances essential to chicken life, feed makers began adding them to chicken feed. Chickens didn't need to go outside anymore. They didn't need to eat cabbage and table scraps or a dead toad to get a "complete" diet. And with those pesky essential nutrients out of the way, it was at last possible to concentrate on the stuff that really made chickens get big fast: carbs. In the late 1940s, a new and important feed was unleashed upon poultrydom: the "high-energy diet." For chickens to grow twice as fast as their recent ancestors, they needed to mainline carbs.

There was, however, a tradeoff that no one thought much about in the 1940s, or today. What the high-energy diet gains in calories, it loses in flavor. The feed is typically a blend of seeds—corn, wheat, millet, soybeans, etc.—and while some seeds (nutmeg, for example) are flavorful, the seeds we feed chickens are not. And unlike tomatoes, a chicken doesn't make its own flavor. The taste of animal flesh is strongly influenced by what an animal eats. Flavor compounds in the food birds eat find their way into bird tissue. Scientists refer to this as biodistribution—it's the same reason a dairy cow that eats onion grass produces milk that tastes like onions.

Flavor-wise, chicken started moving in the wrong direction when the high-energy diet appeared, so much so that Julia Child sounded the alarm in 1961. It's only gotten worse. Never mind that chickens don't eat grass, herbs, or bugs anymore, all the seeds that go into modern broiler feed come from varieties of grain that themselves have been "improved" the same way tomatoes have been "improved." They have experienced massive leaps in yield and corresponding diminishments in flavor. It's hard to imagine a blander diet. Broilers may not be source limited when it comes to nutrients, but they are source limited when it comes to flavor. Like modern tomatoes, modern chickens suffer from a flavor brownout. But unlike tomatoes, it's not because of a lack of nutrition—it's because of an overload of nutrition.

There is a second part to this problem: youth. The high-energy diet, with its dusting of essential vitamins and minerals, enabled the production of giant babies. And meat from babies is bland. Veal is blander than beef. Lamb is blander than mutton. Suckling pig is blander than mature pork (which most people today have never tasted). Part of that is due to water—the younger the muscle, the more moisture it contains. But it is also due to aspects of animal biology that scientists still don't understand, in large part because very few of them are looking.

A COOK named Jeanette Young Norton, who knew not a thing about biodistribution, genetic tradeoffs, and source limitation, nevertheless learned this lesson as early as 1917. That's the year the now totally forgotten *Mrs. Norton's Cook-Book* was published, in whose pages Norton preempted *Mastering the Art of French Cooking* by more than four decades when she thundered, "Of course the birds that are unnaturally mothered, fed, and fattened may make a fine appearance, but the flavor is not up to the real thing." The more ancient *Good Cooking*, first published in 1896, had this to say about how chickens taste: "Full-grown poultry have the best flavor." For fried chicken—which, you will recall, requires a younger, more tender and, therefore, less flavorful chicken— *Good Cooking* suggests a four- or five-month-old bird (quadruple the age of today's broiler). And though this bird was mild by antique standards, it couldn't have been that mild since the recipe suggests seasoning the chicken with salt and pepper and nothing else. No poultry spice, no garlic, no Chicken Tikka Flavour Explosion, no paprika. Just salt and pepper. Same thing for the pan gravy—no sage leaves, no cognac, no bouillon cubes, no instant gravy mix. Just milk, flour, and salt and pepper.

Now consider the fried chicken recipe in a 1902 masterpiece called *The Ideal Cook Book*. It's the same recipe. Salt and pepper. So is the recipe in 1896's *Boston Cooking-School Cook Book*, the *Joy of Cooking* of its day. Nearly every recipe from what we might as well start calling the salt-and-pepper era said that fried chicken needed only two things: salt and pepper.

This was not due to failed herb crops, a trade embargo on spices, or a puritanical preference for plain food. Our forebears were not culinary rubes, as is plainly obvious from the baking section of *The Ideal Cook Book*, whose ninety-three recipes include Indian Bread, hop yeast, and all manner of bygone rolls, waffles, muffins, pone, fritters, and griddlecakes. These folks could cook. They cooked all the time. They made their own biscuits, pickles, and ketchups. (There are eleven ketchups in

The Ideal Cook Book, including cucumber ketchup and grape ketchup.)

It's not like seasonings didn't exist a hundred years ago. Cloves appear ninety-three times in *The Ideal Cook Book*. There is sage in the baked white fish, roast Belgian hare, duck stuffing, beef loaf, and country pork sausages (among numerous other dishes). Thyme shows up in oxtail soup, crab bisque, baked cod, and nut loaf. (*Nut loaf?*) Both the lobster sausages and potted crab require mace. Cinnamon, pepper, and mustard appear more than a hundred times, cayenne sixty times.

Herbs and spices were there if you needed them. But fried chicken didn't need them. Add a little salt and pepper, and chicken took care of the rest.

More than a hundred years later, chicken takes care of nothing. Like a cardboard tomato soaked in dressing, chicken, too, follows the Dorito model. We don't shallow-fry chicken in cast iron pans anymore, we deep-fry it. Now we dip it in buttermilk before dredging it through flour so that it becomes encased in a savory Dorito-like shell. And boy, do we flavor it. Celebrity chef Thomas Keller's 2009 recipe for buttermilk fried chicken is a testament to the current state of teddy-bear stuffing. The brine alone calls for twelve bay leaves, a head of garlic, black pepper, thyme, honey, rosemary, parsley, and five lemons. The batter—just flour back in the day—calls for paprika, garlic powder, cayenne, and onion powder. Southern chef John Currence soaks his chicken for four hours in Coca-Cola mixed with liquid smoke, Tabasco sauce, and Worcestershire sauce, and then flavors the batter with cayenne pepper, garlic powder, and onion powder. These resourceful chefs are fighting more than fifty years of dilution. Flavoring the flavorless is hard work.

Most people don't bother. Chicken might be cheap enough to cook on a Tuesday night, but who has time to brine it for twelve hours? It's easier to buy chicken that's been preflavored. The good news (okay, it's actually not good news) is that there's lots of preflavored chicken.

———————

THE TECHNICAL TERM for preflavoring a chicken is "further processing," and it takes place during every minute of every day in factories that dot the continent. I visited one in an industrial suburb where workers were transferring cardboard boxes from an eighteen-wheeler onto a concrete loading dock. Inside, the frozen contents—uncountable thousands of chicken breasts—were tipped into a giant stainless steel meat grinder along with water and seasoning. This was flavor treatment number one.

The chicken emerged from the grinder as a pungently aromatic gray blob. (Close your eyes and you might think a sixth grader was cooking Italian food for the first time.) It was loaded into plastic bins and wheeled into the next room, where men in white rubber boots used white plastic shovels to feed it into a loud, shuddering machine that pinched the ungainly mass into perfect nuggets and deposited them, one by one, onto a belt that conveyed them on a transformative journey of a few hundred feet. They were coated with batter, which is flavored, and dusted in a breading, which is flavored, and then fried to a golden brown, frozen, bagged, boxed, loaded on another truck, and sent to a national chain of restaurants that appeal to sports fans and families and serves its highly flavored chicken nuggets with a choice of sauces that includes ranch, Thai chili, and BBQ.

Sometimes two patties of ground chicken are glued together into a pouch, and workers wearing bouffant caps and blue gloves that stretch up to their elbows, like ladies' formalwear, gently place frozen pucks of "garlic herb butter" or cheese inside. And not all of the breast meat is ground. Strips are soaked in marinade and tumbled in a stainless steel vat that looks like a giant front-loading washing machine. Then they're battered and breaded, frozen, boxed, and shipped, just like the nuggets. The market is moving away from chicken strips, however, because kids who eat nuggets grow into adults who find strips of actual chicken too chewy. To them, chicken should put up the same kind of fight as popcorn.

This serves everyone just fine. The real money, according to a candid executive at the nugget factory I visited, is made on two things. The first is breading. There is as much breading in a nugget as there is chicken. The second is water. The more water you can add to your nugget, the greater the profit. Further processors, it turns out, are doing to chicken what agriculture is doing to fruits and vegetables: adding water and carbs.

When you stop to consider that nearly half of all chicken sold is "further processed"—chicken nuggets, chicken sausage, chicken patties, chicken burgers, chicken strips, chicken cutlets, chicken Kiev—that adds up to a lot of preflavoring. If you have visions, as I did, of a man wearing a chef's hat sampling a nugget and then sprinting to the other end of the factory while shouting, "More oregano! More oregano!" think again. Chicken factories think about flavoring the way they think about toilet paper: Don't run out. The seasoning arrives by the truckload and is stored in big paper sacks in a kind of flavor warehouse alongside sacks of powdered marinade, breading, and batter. Just tear open a bag and dump it in.

It turns out flavor isn't the restaurant's concern, either. Chicken flavoring is a whole industry of its own. Those brown sacks all bore the name of a company that wasn't the nugget factory, wasn't the restaurant, and wasn't anyone else involved in the hatching, raising, dispatching, and cooking of chickens. The company is called Griffiths Laboratories, and it is just one among many firms, some very big and others small and extremely specialized, that are in the business of providing what are known as "flavor solutions."

Because however complicated it may be to breed and grow chickens, pigs, and cows, and bumper crops of tomatoes, corn, and lettuce, that's actually the easy part. Making a modern chicken taste good requires a flavor solution that calls for three rounds of seasoning that includes recognizable substances like garlic and oregano, unrecognizable substances like MSG or hydrolyzed yeast, and unknowable, secret substances called "natural flavors" or "artificial flavors." Making flavor solutions requires a cluster of scientific know-how that brings together advanced organic, analytic, and synthetic chemistry with engineering, neuroscience, psychography, psychophysics, ethnography, demography, molecular biology, finance, botany, economics, and physiology—even feelings.

If you are wondering why you've never noticed this thing called the dilution effect, why all that supposedly bland food we now grow tastes delicious—why people still want it despite its nutritional and flavor diminishment—it is thanks to flavor solutions. So much of the food we now eat is not only a lie, it is a very good lie. Modern food may be the most compelling lie humans have ever told.

THREE

Big Flavor

THE FOURTH time Hank Kaestner visited Madagascar, he got stuck in an elevator. Kaestner always visited in spring, he always stayed at the Hilton, and over those four visits, the hotel elevator performed very much like the nation's economy. The first three times, it ran smoothly; now it was stuck. To lovers of milk shakes, chocolate, ice cream, and cake frosting, this was grave news.

Kaestner was a spice buyer for McCormick & Company, a profession as glamorous as spying but with better food. One week, Kaestner might fly to Brazil to buy a few tons of cloves, the next he'd go on an allspice expedition in the Mexican jungle and get swarmed by killer bees. He once held a private meeting with the king of Tonga. Kaestner loved his job for two reasons: He loved spices and he loved the amazing places where spices grow. And among all those places, Madagascar was special because of what grew there: vanilla, which Kaestner calls "the most magical spice."

And now Madagascar was in trouble. The problems started in February 1975, when the president of Madagascar was assassinated. A few months later, the country's new leader, a military man named Didier Ratsiraka, nationalized banks and industries and declared the country a Marxist republic. Within months, embassies were closed, and tourism shriveled. At the formerly deluxe Hilton, there were burned-out bulbs in the lobby and North Koreans, brought in to do security, walking the hallways.

Vanilla production was devastated. By 1976, Madagascar was producing less than half as much vanilla as it had the year before, and in 1979 it was down to a third. Back in his office at McCormick headquarters in Hunt Valley, Maryland, Kaestner received a package containing photos of a great swath of cured vanilla beans being crushed by a steamroller. The government was destroying its buffer stock. They wanted the price to go even higher. It doubled.

Needless to say, this was distressing news for McCormick. Vanilla was a major profit center. The company could make as much money off a few hundred tons of vanilla beans as it could from ten thousand tons of black pepper. The reason was its flavor. Everyone loves vanilla. Nothing improves cake frosting or lifts French toast like a few dark drops of pure vanilla extract. Vanilla is flavoring royalty. It's better than caramel, better than almond, and better than toffee.

Everything about vanilla extract was perfect except for one thing: the price. Even under non-Marxist republic conditions, the stuff was expensive. Making it requires cultivating vanilla orchids, pollinating the blossoms by hand, waiting for the beans to ripen, picking them at just the right time, boiling them in water, "sweating" them in hot chests or tanks, setting them out every morning to bake in the sun until they're dried, and then conditioning them in closed boxes for a period lasting months, at which point the beans—which are now as moist as a raisin and as long and dark as a cigarillo—are shipped to Europe, then to New York, and then to an extraction plant, where they're chopped into tiny pieces and alcohol is passed over them continuously in a process of steeping that can take more than a day, and finally held for weeks so the brew can "settle." It takes a year and half to go from orchid blossom to extract, a single ounce of which costs as much as a shot of good single-malt whiskey. (A bargain.)

And now there was a dark cloud hanging over vanilla. Prices were going up. Supply was dwindling. A home gourmet or a high-end pastry chef might tolerate expensive extract, but McCormick had customers who manufactured ice cream, yogurt, beverages, chocolate, and pastries, and who ordered vanilla extract by the gallon. What were they going to do?

Two years after Hank Kaestner got trapped on that elevator, McCormick asked a question: Is there an easier way to make vanilla? There was. And the answer, though spectacularly complex, was also ingeniously simple: fool people.

ONE HUNDRED and seven years earlier, a German chemist at the University of Berlin asked the same question. Wilhelm Haarmann possessed an exceedingly strange interest: pinecones. He believed they were hiding a secret: the potential to produce an almost magical white powder that could make pastries, drinks, and chocolate taste better.

The powder itself was no secret. It had been discovered years earlier by a Frenchman who purified and filtered vanilla extract until he was left with a white crystalline substance that smelled potently of vanilla. The new substance became known as "vanillin" (pronounced VAN-illin). But then, for nearly two decades, nothing. Vanilla's secret wasn't a secret anymore, but no one could do much about it because vanillin could be made only from vanilla extract. And since vanilla extract was already expensive, vanillin was said to be worth more than its weight in gold.

Haarmann, however, knew something others didn't. Years earlier, in Haarmann's hometown of Holzminden, a pharmacist had been experimenting with a substance he scraped from the inner bark of pine trees. He filtered, pressed, boiled down, and purified the gooey material until he was left with crystals he described as "white, silky-sheened, very delicate." When the pharmacist squirted acid on these crystals, an extraordinary reaction took place. The air became perfumed with vanilla.

Was it possible to produce this precious and exotic tropical substance from something as ordinary and abundant as pine trees? Haarmann managed to get his hands on the old pharmacist's remaining stash of pine crystals and, in his laboratory, pulled off a chemistry miracle. He turned pine crystals into vanillin. It was as though he had created a key that fit the lock of a door that had never before been opened.

In 1875, after collecting forty-five pounds of pinecones in the Black Forest, Haarmann opened Haarmann's Vanillinfabrik. What had formerly been the exclusive domain of a tropical orchid was now being produced in a factory in Germany. Pinecones went in one end, vanillin came out the other. Haarmann's company would eventually figure out how make vanillin from clove oil, which was even cheaper, and would go on to manufacture a synthetic flavoring found in violets, which is still used to make fruit flavors. His flavor business would become so successful that Holzminden would become known as the City of Fragrances.

THAT PROBABLY sounds odd. A city that makes flavors becoming famous for fragrances. That's because so much of what we think of as flavor is actually aroma—scent, bouquet, odor, and so forth. So much of what characterizes the food we love is no more substantial than perfume. And it is explained by a phenomenon known as "retronasal olfaction," a fancy term for back-of-the-nose smelling. Retronasal olfaction happens when an aroma enters your nose not through your nostrils but through that hole in the back of your throat. It is a fundamentally different kind of smelling from nostril smelling, engaging more and different parts of the brain. Of all the senses, it is both the most intense and, curiously, the most unknown. We all know we can see, hear, feel, taste, and smell objects in front of us. Almost no one has any idea that the nose's greatest talent is appreciating food. (The people who come closest are wine aficionados, who exhale through their nose to fully appreciate the flavor of wine.)

This is how it works: When you eat food, the combination of saliva, pressure, and heat releases "volatile aromatic compounds"—wafting food vapors, basically, like the ones you see emanating from burbling pots and grilling steaks in vintage cartoons. As you eat, these aromatic gases float up or are exhaled into a dank, snot-lined chamber inside your nose, called the nasal cavity, situated directly above the roof of your mouth, the ceiling of which is dappled with microscopic pouches that catch odors.

These pouches are so tiny that whole molecules don't fit inside them, only parts of molecules. When a receptor is "stimulated," it sends a signal through the human body's most direct conduit of nerve fiber to the olfactory bulb, an ancient olive-shaped piece of brain that hovers over the nasal cavity as though it, too, is trying to get in on the smelling action. And if you think that all sounds complicated, hold on tight. A single molecule can have lots of different parts and, therefore, stimulate many different receptors. A single receptor, similarly, can be stimulated, to varying degree, by different kinds of molecules. There are around four hundred different kinds of smell receptors, but the combination of signals—each combination represents a different aroma—is, at present, incalculably huge. But we do know this much: We are capable of distinguishing more than a trillion aromas.

The brain, fortunately, is expertly talented at cataloging them. The aromatic compounds that waft off bacon, for example, trigger a combination of smell receptors that a person remembers as "bacon." When that person walks into a diner at 9 a.m. and is struck by the scent wafting off the sizzling flattop and a very similar combination of receptors is fired, it triggers an experience that is as much analytical as it is emotional: "bacon."

When it comes to sensing food aromas, the human tongue is as useful as the human big toe. If you plug your nose and dump an ounce of vanillin on your tongue, you will taste only a very mild bitter flavor. It's not until you inhale through your mouth and then exhale through your nose and vanillin molecules reach the nasal cavity that the silky tropical scent tingles your brain.

The tongue is far from useless, however. The tongue senses what are called the "basic tastes": sweet, salty, sour, bitter, and umami (savory). Actually, it isn't just the tongue. There are taste receptors on the roof of your mouth and the back of your throat. Sweet things, like sugar, stimulate sweet taste receptors, which produce the universal, hardwired pleasure known as sweetness. Acids stimulate sour receptors. And that's how it works for bitter, salt, and umami.

The flavor industry has not forgotten about the tongue. The discovery of umami rivals vanillin for its influence on the food industry. That happened in 1908, when a Japanese chemist named Kikunae Ikeda wondered what it was that gave foods like cheese, meat, and Japanese fish broth their irresistible savory quality. After manipulating seaweed with the cutting-edge chemistry of his day, he discovered fish broth's inner secret: a crystalline substance called glutamic acid. (This was the era of exciting crystals.) The next year, Ikeda began selling an edible form of glutamic acid known as monosodium glutamate under the brand name Aji-no-moto. Today, Ajinomoto is the world's leading producer of MSG (not to mention a whole lot of sweeteners), and it's worth around five times as much as the company Wilhelm Haarmann founded, which is now called Symrise.

It might be conceptually helpful to separate "taste" and "aroma" to understand how flavor works, but it is the combination of the two that we find so stirring. Without the mouth, food is nothing more than a fleeting scent. And without the nose, it's dismally simple. When a person eats bacon, receptors in the mouth sense saltiness, sweetness, and umami while the nose senses its sweet, roasty, smoky, porky volatiles. In the mind, they combine to form a blend that is vivid and inseparable—and deeply pleasurable. Bacon.

BY THE mid-1970s, the world was awash in Wilhelm Haarmann's wonder powder. Vanillin was cheap, it stored easily, and it didn't come from politically unstable former French colonies. Everything about vanillin was perfect except for one thing: quality.

Vanillin may have been vanilla's biggest secret, but it was by no means its only secret. Vanilla contains hundreds of other aromatic compounds. Not a single one of these comes anywhere close to the dominance or likability of vanillin, and some of them don't smell very vanilla-like on their own. These notes include "woody," "rummy," "smoky," and "watermelon." They are, nevertheless, essential to the experience of authentic vanilla. They give it what flavor scientists refer to as "depth," "structure," "body," and "dimension." On its own, vanillin is a blast of sweet cotton candy—fun but simple, a good-looking dimwit. If vanilla is a densely layered classic novel, vanillin is a comic book.

By the late 1970s, with Madagascar now in the business of destroying its own vanilla beans, vanillin, however simple, was starting to look like a pretty good option. For McCormick, the situation was bad. McCormick wasn't much of a flavor company. It was mainly an herb and spice company, which it still is today. It dabbled here and there with chemical flavorings, but it was a pipsqueak compared to the multinational flavor behemoths like Givaudan, Haarmann & Reimer, and International Flavor & Fragrances that could pump out huge quantities of vanillin. Vanillin was a commodity. The real money was in true vanilla extract. The real money was in complexity. And that's what the company set out to achieve. In 1978, McCormick set its sights on "complex" fake vanilla.

The team featured a young food technologist named Marianne Gillette with a master's in nutrition from UC Davis. In one room, Gillette assembled a "trained descriptive panel" of expert sniffers and tasters who specialize in identifying flavor notes in food. They sniffed and tasted pure vanilla extract and scribbled down notes like "vanillin," "woody," "pruny," and "rummy." Gillette then took the notes to a flavorist—a professional concoctor of flavor chemicals—who pulled out bottles filled with the exact synthetic matches of "vanillin," "woody," and "rummy," and blended them. A day later, the sensory panel would convene again. Gillette would have them compare the latest fake vanilla with pure extract, and the sensory panel would proceed to rip into it. "Too much vanillin," they might say, or "I'm not getting the woody note."

After two years, the flavorist's artificial vanilla started smelling like a pretty good match for real vanilla. But something was still missing. There was a gaping aromatic hole in the middle of fake vanilla. The professional sniffers called it "resinous." Gillette came back to the flavorist and said, "There's not enough resinous." And here the flavorist drew a blank. There was no match for "resinous" in her arsenal of flavor chemicals. Just what the heck was "resinous," anyway? The panel described it as "leathery."

It was time for gas chromatography.

This ominous pair of words is the name of the technology that blew the flavor door off its hinges in the 1950s. Before gas chromatography, there was no good way to separate out, for example, the multitude of aromas in vanilla (or any other food, for that matter). Beyond vanillin and perhaps a few others, flavor scientists could only stare in wonder at a bottle of extract. With gas chromatography (GC), suddenly you could put a drop of extract in a machine, and over the next couple of hours the individual compounds would come marching out the other end, one by one. The machine even produced a printout that displayed each chemical as a peak. A big peak meant there was a lot of that chemical and a tiny peak meant there wasn't much at all.

By the time of the resinous mystery, McCormick had already analyzed the bejesus out of vanilla and used the results to create a chemical "map" of the flavors inside. But the sensory panel's resinous finding was a revelation nonetheless. The map was incomplete. Somewhere in vanilla's chromatogram, there was a small peak everyone had been missing.

This time, instead of scouring the printout for peaks, the flavorist pulled up a stool next to the GC. As each compound came out the other end—woody, beeswax, rummy, smoky—she sniffed, careful not to burn her nose on the hot gas, hoping resinous would eventually appear. For nearly an hour, there wasn't so much as a peep. Then, a fleeting, leathery wisp. Resinous. They checked the printout. The peak was so small it looked more like a valley.

But for resinous, the jig was up. Its decades of secrecy were at an end. A McCormick chemist named Patrick Hoffman cranked the sensitivity on the GC way up and ran the vanilla profile over and over. He got different flavorists—some male, some female—to sniff the output. The suspect was apprehended. Once resinous had been captured in a bottle, McCormick identified its chemical makeup using a different technology called mass spectrometry, which deduces unknown substances by measuring their molecular weight. (It's like trying to figure out what's inside a mysterious piece of luggage by heaving it off a hotel balcony.)

And boy, was resinous ever potent. The chemical, which, more than thirty years later is still a closely guarded trade secret, registered in parts per trillion. According to Hoffman, who would go on to become McCormick's director of technical sciences and is now retired, it would take just ninety-six grams of resinous (a little over three ounces) to give Niagara Falls a discernible leathery note for an hour.

It wasn't long before McCormick was in the resinous business. In 1982, four years after Gillette and her colleagues first set out to crack the vanilla secret and six years after Hank Kaestner got stuck in that elevator, McCormick began selling a sweet-smelling liquid called Imitation Vanilla. Whereas vanilla extract featured hundreds of compounds, Imitation Vanilla had around thirty, and not one of them was made from vanilla beans. Vanilla had been chemically decapitated and its production outsourced from Madagascar to suburban Baltimore. Only the chemists knew about it. To a person licking an ice cream cone, it tasted exactly the same. According to Gillette and every member of her sensory panel, it was a "perfect match" to pure Madagascar vanilla extract. Industry agreed. Many customers switched to Imitation Vanilla and never switched back, even when the price of real vanilla came back down.

A century after Wilhelm Haarmann stepped into that German forest, a fantasy that began with pinecones was realized. Fake vanilla was cheap and available by the gallon. And it didn't taste fake anymore.

AND THAT, ladies and gentlemen, is how industry has accomplished what you might call the great flavor migration. One after another, humans have captured the chemicals that characterize foods like apples, cherries, carrots, and beef and moved their production from plants and animals to factories. In 1965, there were less than 700 of these chemicals. Today, there are more than 2,200. The most recent additions are "mustard/horseradish/wasabi" (2-(methyl thio) ethyl acetate), "raw potato" (2-ethoxy-3-ethylpyrazine), and "balsamic vinegar" (Sclareol). But don't think of it as a list. It's more like a vocabulary, because these chemicals are mixed and blended in an almost endless combination to produce knockoffs that keep getting more complex, layered, and convincing.

None of it would have happened without gas chromatography. The first commercial GC unit went on sale in 1955, and soon deltadodecalactone, which had been found in butter, was being added to margarine. A "green note" called leaf alcohol paved the way to better fake strawberry, a flavor that got even more convincing in 1964 with a sweet-smelling substance called Furaneol. That same year, nootkatone was discovered in grapefruits, and by 1970 the Coca-Cola Company was adding it to Fresca.

The age of human-made flavor had dawned. It was now possible to fool people into experiencing mintiness without a single leaf of mint. You could create the vivid sensation of passion fruit with just a few drops of liquid. In 1986, the chemical secret of the hazelnuts was out of the bag, and five years after that white truffles lost their historic monopoly on 2,4,6-trithiaheptane (the "vanillin" of northern Italy's rarest and most expensive ingredient). For thousands of years, the flavor of orange could be experienced only via the seasonal, perishable spheres of fiber, juice, vitamins, minerals, and antioxidants called oranges. Now you could order "orange" by phone and add it to soft drinks, popsicles, yogurt, or gum. Name any "thing"—raspberry, chicken, pineapple, tomato, blueberries, even tacos—and there were chemicals that imitated it.

Not all flavor breakthroughs involved the nose. By the late 1960s, the umami arsenal included hydrolyzed protein, autolyzed yeast, and torula yeast (which was in Arch West's Taco Doritos), along with disodium inosinate and disodium guanylate, which amplify the sensation of umami. In the early 1990s, scientists at Ajinomoto (the MSG company) turned their attention to garlic and wondered, in a Haarmannesque moment of curiosity, what it was about garlic that made it so universally liked. They were thinking: aroma. The onion family (garlic is a type of onion) is notorious for pungent sulfur aromas. When they pulled garlic apart into all its chemical pieces, however, the magic ingredient, they claim, turned out to be something else: a protein known as glutathione. This wasn't a complete surprise, at least not to Ajinomoto. Company scientists had already found glutathione in fermented soy products like miso and soy sauce and suspected it was contributing to the deep, rounding, and satisfying effect they had on food.

When glutathione was added to beef broth, a panel of fifteen trained sensory experts (the flavor industry is full of trained sensory experts) found that it "strengthened" the meaty flavor and added "continuity," "mouthfulness," and "thickness." But strangely, glutathione only ever did something in conjunction with sweet, salty, and umami tastes—by itself, it was tasteless. Ajinomoto christened this newfound sensory effect as kokumi, the Japanese word for "rich tasting." In a later study, Ajinomoto even found what would seem to be a kokumi receptor on the tongue. Some people have now taken to calling kokumi the sixth basic taste, but not everyone agrees—how can you call something a taste if you can't taste it?

Everyone is eating kokumi. They were eating it before it even had a name, when it was part of soy sauce, garlic, and meat. But now they're eating more of it because Ajinomoto is cranking out kokumi flavorings in dump truck quantities. The world's largest flavor company, Givaudan, is also in on the kokumi action, which it promotionally describes as "full bodied," "hearty," and "rounded." If you have eaten chicken wings in a restaurant lately, or potato chips, broths, gravies, or any other savory snack, there is a good chance a billion-dollar multinational you've never heard of has stimulated your still controversial kokumi receptor.

THE WOMAN who helped crack the resinous mystery was having me tick boxes. Happy—check. Good-natured—check. Warm—check. Loving—check. I was standing next to Marianne Gillette in the sensory lab at McCormick's Technical Innovation Center. I had come to the same complex of buildings in Hunt Valley, Maryland, where "resinous" had been captured and Hank Kaestner received the photos of steamrolled vanilla beans. I had, a moment earlier, just finished sampling an applesauce flavored with the best cinnamon I'd ever tasted. It was called Saigon Cinnamon (it comes from Vietnam). Next to it was a bowl of applesauce made with Indonesian cinnamon, and it tasted like someone had stirred in a spoonful of dirt.

Now I was rating cinnamon's "emotional profile." The very idea struck me as touchy-feely, pseudo-empathetic, focus-group-style garbage. But as I basked in a Saigon cinnamon afterglow, the answers were coming faster than I could put ticks on the page. I felt "pleasant." I felt "pleased" and "peaceful." I did not feel "guilty," "worried," or "disgusted." Cinnamon, I realized, is the flavor equivalent of being hugged by your grandmother.

Gillette showed me McCormick's EsSence Profile for cinnamon. The whole world, apparently, feels happy, glad, affectionate, and understanding. No one feels wild, daring, or bored. As I came to grips with the unoriginality of my views on cinnamon, Gillette gestured to the profile and said, "These are remarkably consistent."

Gillette is now McCormick's VP of applied research. Over a career spanning nearly four decades, she has devised a method of making sure that fast food always tastes the same and has cowritten academic papers with titles like "How Much Pain Will Bring You Pleasure?," "A Method for Testing the Ease of Extrusion of Icing Marketed in Plastic Tubes," and "The Demographics of Neophobia in a Large Commercial US Sample." A "neophobe" is someone who doesn't like trying new foods. Gillette is not one. She has personally tasted and evaluated more than 6,300 different types of commercial food products. Vanilla extract is just one of those products, and she's sampled it around 2,000 times. Her long hair is grayer now than it was during the resinous mystery but her smile is as bright as ever. If you passed her walking down the street you might think she is a professor of something liberal sounding, which isn't far off.

Flavors are about feelings for Marianne Gillette. Feelings matter. "If you were creating a cracker for a spa," Gillette explained, "you'd go for something that's whole, understanding"—she ticked off each word with care, as though composing a personal ad—"tame, steady, mild"—she paused—"and calm." But if you were creating a cracker for a tailgate party or a sports bar, Gillette continued, "you'd want seasoning that was wild, active, adventurous, daring, eager, and enthusiastic."

The reason crackers sound like personal ads is that space in the cracker aisle is limited, so the cracker companies get only one shot. The companies that make crackers—or burgers, chicken nuggets, snacks, drinks, or any other food product—are perpetually afflicted by the same anxiety as men at a singles' bar. They do not understand what their love interest wants.

McCormick can help. The company that most people associate with their grandmother's spice rack knows exactly what grandmothers with spice racks look for in a tea biscuit. (Butter, Gillette says.) For example, a national chain restaurant once approached McCormick because it wasn't getting the kind of fajitas sell-through it expected. When Gillette and her colleagues visited the restaurant, they observed the ritual of the fajita moment: An awe-struck silence would sweep across the dining room as a waiter carried a sizzling fajita skillet to some lucky table. They went back to the office and brainstormed. How can we make this moment even more dramatic? They created a "sizzle sauce," which made the sizzle louder and the aroma more intense. Sales spiked.

McCormick once made a cedar-plank flavor for a restaurant that didn't want the bother of cooking salmon on actual cedar planks. Using the same technology it used to create the imitation vanilla, McCormick has created "Ultimate Lemon," which was formulated using aroma chemicals found in lemon peel, Meyer lemon, lemon thyme, and Limoncello (a refreshing and highly drinkable Italian liqueur). Ultimate Lemon might show up in a beverage, dessert, or salad dressing. Not that you'll ever know. Whether it says so on the label or not—and it usually does not—McCormick is in every aisle and on every shelf of the supermarket. The company provides "custom flavor solutions" for nine of the top ten American food companies and eight of the top ten food service companies. (Food service refers to large chain restaurants, companies that sell to smaller restaurants, school cafeterias, hospitals, and so forth.) McCormick is in your pantry, your fridge, your freezer, and nearly every restaurant. Unless you are a hunter-gatherer or have spent your life obtaining calories via feeding tube, McCormick has used the science and psychology of food to make you happy. It's probably happened in the last week.

Flavor solutions are born in the Create It Center where Gillette and a colleague named Suzanne Roy, who has dark hair and a slight French Canadian accent, walked me through the exploding world of crackers. Just a few years ago, crackers were like tortilla chips circa 1962—"uni-dimensional." People would come home from work and put stuff—pâté, a piece of cheese, a pickled pepper—on a cracker and eat it. Ho hum. Then, Roy explained, herbs got big. Herbed crackers were tastier than plain ones. People started eating them right out of the box. They didn't have to bother spreading pâté, slicing cheese, or fishing a pickled pepper out of its brine. Now crackers are moving into the next frontier—herb with something else. Herb and garlic. Herb and olive oil. It isn't going to stop there. A new and previously undiscovered galaxy of flavored crackers is unfolding. "I could see smoked paprika, pancetta, and cara-melized onion," Roy said, crystal-balling. "If a cracker is more flavored," she added, channeling the spirit of Arch West, "you don't want to put something on it. They're made to stand alone."

When you see how McCormick goes about thinking up a new cracker flavor, it becomes apparent why other companies find this all so chal-lenging. One thing you don't do, for example, is just sit down and think about crackers. You look at what's hot in the drinks segment. You look at what's happening with children's toys, technology, fashion, or the lat-est color palettes featured in home décor magazines. You even look at cars. (In 2010, McCormick pushed its pumpkin pie spice mix after not-ing pumpkin was a big color with automakers. Sales spiked.) You look at the hot new food trends in Asia, Australia, Brazil, or anywhere else a trend is breaking.

Above all, you watch people. If a national chain of family friendly sit-down restaurants asks McCormick to develop a pork sandwich, Mc-Cormick will go to that restaurant. What does it smell like? Who's walk-ing through the door? Did they drive? What kind of car was it? What are they wearing? Who are they with? Is it a date? Are they with friends? What time is it? What did they order? What are they *feeling*? McCormick visits other places that sell pork sandwiches and pays very close attention to people in search of a flavor experience that will make them feel adventurous, satisfied, aggressive, and secure. McCormick watches people everywhere—convenience stores, cafeterias, hotels, airports, schools.

They observe "need states." A need state is an emotional requirement for food. A single mom who has to put together dinner for her fussy teenage daughters has a need state for a no-fuss dinner with a hint of adventure—the flavors of an exotic Asian country with lush forests and waterfalls, perhaps—to shoo away the dreariness of the Tuesday evening meal. She wants food that makes her family feel steady, satisfied, and pleased, but also enthusiastic, energetic, and a little bit daring. An unmarried male software coder in his early twenties grabbing a bite before a six-hour multiplayer session of *Guild Wars 2* might share the identical need state. So might a sixty-eight-year-old retiree in Arizona who played golf all afternoon but promised his wife he'd fix supper. McCormick used to see the world in demographics—moms, single males in their twenties, men over sixty-five. Need states are a more nuanced view of the human condition. That paprika, pancetta, and caramelized onion cracker may not exist yet, but when it does it will cut across race, gender, and age. We are united by an emotional need for flavor.

The tricky part is predicting the objects of these needs. They keep changing. That's the reason there's a multibillion-dollar sector of the food business known as food product development. These companies think about flavors the way quarterbacks think about receivers: It's not where flavor is; it's where flavor is going to be. Every January, McCormick releases what it calls its Flavor Forecast, which is the industrial food world's equivalent to *Vogue*'s September fashion issue. The 2003 forecast announced a soon-to-be-popular Mexican smoked pepper that went triple platinum: chipotle. Numerous predictions have revealed an astonishing level of insight, including sea salt, pomegranate, smoked paprika, and coconut water. But the best prediction McCormick has made is the prediction that their predictions would be considered predictive and important. When the Flavor Forecast launched in 2000, no one else predicted flavors. Now everyone is doing it.

Gillette brought me down for lunch to McCormick's Culinary Conference Room, where executive chef Kevan Vetter prepared a meal of new flavors that would become flavor solutions and satisfy need states over the next year or two. As we ate mustard grilled brown sugar bourbon pork tenderloin, Korean grilled chicken wings, and roasted baby broccoli seasoned with a Moroccan spice blend called dukkah (hazelnuts, cumin, sesame seeds, coriander, salt and pepper), Gillette and her colleagues recounted the flavors of times past. Like hipsters referencing obscure hair bands, they recounted the must-eats of the '80s: pesto, salsa, yakitori, Cajun, passion fruit, sun-dried tomatoes. Hot sauces rumbled in the late '80s and erupted in the '90s, when "big and bold" became the industry catchphrase. That was the decade adobo, miso, Tex-Mex, and goat cheese went mainstream and celebrity chef Emeril Lagasse implored the country to "kick it up a notch."

That was many notches ago. Chipotle. Chai. Pomegranate. Açaí. Every few years, America rouses from its food coma and demands excitement. Like crackers, food trends are becoming "multidimensional." In 2007, McCormick started pairing flavors—wasabi and maple; cumin and apricot; clove and green apple. In 2012, combinations of three or four appeared. Rosemary, which made a quaint solo appearance as "rosemary" in the 2000 Flavor Forecast, returned in 2013 as part of "smoked tomato, rosemary, chile peppers & sweet onion." The pace is picking up. Flavor trends are becoming like tropical weather. They roll in, they're intense, and then they're a memory.

All of which may be one reason so few people cook at home these days. Who can keep up?

Restaurants can. If McCormick determines that what that national chain of family friendly sit-down restaurants really needs is a cinnamon-chipotle pulled-pork sandwich, it might take McCormick a hundred iterations to get everything right—the perfect level of spice, smokiness, and no cinnamon burn. At that point, McCormick figures out how such a sandwich can be produced on a mass scale.

Because the restaurant sure isn't going to make it. Most restaurant chefs are "chefs" the way someone assembling an IKEA bookshelf is a "carpenter." And the restaurants in which they "cook" are just pretending to be restaurants. That big tiled room the waiter calls the kitchen is more like a prep area. Dishes aren't prepared so much as assembled by staff that follows "operational guidelines," not recipes. It all involves a great deal of reheating.

The cooking part actually takes place at a "processor" like that chicken nugget factory I visited. There, the pork will be marinated, seasoned, cooked, and pulled, then frozen and sent to a distribution center (a warehouse full of cooked frozen food) and, later, to a "restaurant," where it's "rethermalized," likely "top-noted" with sauce from another processor, and then placed on a bun originating from yet another factory, which arrived partly baked and frozen and was finished in an oven and listed on the menu as "freshly baked." It's not only fast-food restaurants that do this. Chain restaurants do it. So do independent restaurants, bars, pubs, diners, cafeterias, schools, hospitals, retirement homes, and coffee shops. Customers order off menus. The kitchens also order off menus. The real kitchens are factories. And the real cooks are companies like Griffith Laboratories, Kerry, and McCormick.

THE FOURTH FLOOR of McCormick's Technical Innovation Center is where need states are quenched by chemicals. That's where the flavorists work. One of them, a woman with brown hair in a white lab coat wearing protective eyewear, showed me how it's done.

She was making an imitation cinnamon. She started with the primary base note—the vanillin of cinnamon, if you will—a gin-clear liquid called cinnamic aldehyde that smells like cinnamon with too much cosmetic surgery: impressive to a teenage boy, but exaggerated and lacking depth. She dipped a long instrument called a pipette into various top-secret clear liquids—"character notes"—and began adding drop after drop of complexity to the cinnamic aldehyde. One note smelled like tobacco, then came a woody scent, then a citrus note, and last a hint of a something between menthol and eucalyptus. The result wasn't quite as rounded or nuanced as actual cinnamon, but it was way cheaper, and smelling it still made me feel glad, affectionate, and understanding. Without real cinnamon to compare it to, I would never have called it out as a fake.

That beaker of imitation cinnamon was many things: potent, convincing, inexpensive, plentiful. But also "natural." If it were used to flavor a granola bar or a package of instant apple and cinnamon porridge, the back of the package would list it as "natural flavors." How, you might wonder, can a blend of pure chemicals created in a lab by a scientist with a graduate degree who works beneath a vent hood be called "natural"?

The reason is that just as vanillin could be made from pinecones, other flavor compounds—like those found in apples, strawberries, carrots, and so forth—pop up all over nature, very often in cheap sources like leaves, roots, and yeast. Many, many chemical reactions ago in factories far, far away, each of these chemicals was derived using "natural" chemical processes, like organic solvents, enzymes, or heat, and purified using more "natural" processes, such as distillation.

It is not only possible but also much easier to make flavor chemicals in a non-"natural" way with petrochemicals and industrial chemistry. But when you make flavors this way, they must be listed as "artificial." The actual molecules sitting in the beaker are, in many cases, exactly the same whether it's artificial or "natural." The word "natural," in fact, has nothing to do with the end product. It simply refers to the process that gets you there. If it seems silly, that's because it is silly. It's like saying if you walked over the Golden Gate Bridge you'd arrive in "Natural San Francisco," but if you took a cab, then legally you'd have to tell people you were in "Artificial San Francisco." San Francisco is San Francisco whether you get there by birchbark canoe or jet pack. The same goes for vanillin, cinnamic aldehyde, and the gang. For this reason, flavor chemists often think the whole idea of "natural" flavors is ridiculous. But to a concerned parent buying soy milk or a man who knows much more about Bikram yoga than synthetic chemistry, the word "natural" brings to mind wholesome forests and nourishing orchards. For this reason, the word "natural" is very important to the folks down in marketing.

(I will henceforth use the term "synthetic flavor" to refer to flavors engineered by humans and "real flavor" for flavor experienced in the context of a plant or animal.)

Now it was my turn. The flavorist handed me a beaker filled with a flavor that smelled like strawberry gum from the leaded-gas era and invited me to improve it. Unlike cinnamon and vanilla, strawberry has no major characterizing compound. It's a blend of nearly four hundred aroma chemicals, none of which smells like strawberry on its own. I was handed a pipette and had a choice of three of these flavor chemicals, all of them clear liquids in small glass containers. One smelled like burnt sugar and another like berries. The third was special. It smelled vividly green and quenching, like carrot juice served in heaven. On second sniff, it evoked a freshly mowed perfect lawn. The intensity was astonishing. I kept dipping my nose to the rim to pull in another whiff. The chemical is called cis-3-hexenol, which is the fancy name for leaf alcohol, the flavoring that did so much to improve strawberry flavors back in the early '60s.

"Where do you get this stuff?" I asked, thinking about ordering a case.

The flavorist handed me a book called *Allured's Flavor and Fragrance Materials*, which is something like the phone book of flavors. I found cis-3-hexenol on page 187 and it didn't seem special and exotic for very long. A company called Advanced Biotech makes it. So do A. M. Todd, De Monchy, Eramex, and Natural Advantage. All together, fifty-seven companies crank out gallon upon brilliant gallon.

These firms are the plankton of the man-made flavor web. Flavor chemicals are born in ordinary vats inside nondescript buildings just off the interstate and sold to companies farther up the flavor web— Givaudan, International Flavor & Fragrances, Symrise, Firmenich, or perhaps even McCormick (which, in the world of synthetic flavorings, is a small fish). They buy it and precisely blend it, drop by drop, with other flavor chemicals until it bears a stunning resemblance to something real, even though there's none of that real thing in it. Eventually, that high-precision blend winds up at a food processor, where it's added to food—chicken, pork, yogurt, potato chips, fruit drink, candy. From there it's a series of truck rides to a restaurant or supermarket, and then your mouth and lungs before the last little push to your nasal cavity and its aroma receptors, when the journey is at last complete.

IN 1918, the average American sprinkled half a pound of spices on his or her meals over the course of a year. We know this because that's the year the Department of Agriculture began keeping track of spice consumption. The year the Taco Dorito was launched, the number had tripled to one and a half pounds. By 1997, it hit three pounds, and today it's at three and a half. It is an astonishing increase—500 percent. A hundred years ago, the average American could carry a year's supply of spices in one hand. Today, she'd need a bucket.

No one kept track of flavor chemicals in 1918 because there wasn't much to keep track of. The industry was in its infancy, selling things like fruit syrups, spice extracts, and a handful of laboratory creations, such as vanillin. Today, synthetic flavors have infiltrated nearly all restaurants and every aisle of the supermarket. Today, there are chemicals for every need state.

"Industrial" foods like chicken nuggets and rethermalized pork sandwiches feature flavorings. But so do all sorts of "real" foods. Tyson's "All Natural" young chicken is just a raw chicken as plump and bare as a newborn babe. There's no bold southwest flavor, no chipotle, no Asian theme to speak of on the packaging. And yet it, too, receives the Dorito treatment—the ingredients lists "up to 12% chicken broth," and that broth contains "natural" flavorings. So does Perdue's "Tender & Tasty" whole chicken. The flavor of that chicken is partly due to the dead bird wrapped in plastic, but also due to an unknown flavorist in a lab somewhere. Even raw chicken has been engineered to seduce the senses.

It's not just chicken. Smithfield's pork shoulder roast, a hunk of raw pink muscle, contains pork broth and "natural flavoring." Hormel's "Always Tender Original Pork Loin Filet" (also raw) is perhaps less original than its package suggests because it contains "up to thirty percent of a patented flavoring solution" that includes yeast extract, natural flavor, and pork broth as well as sodium diacetate (a white crystalline powder that tastes like vinegar). Even steak, bloodred and just itching to hop on the grill, arrives at supermarkets listing the following ingredients: "beef and natural flavorings."

The salt-and-pepper era feels as distant as the Precambrian. Chicken Rub, which is part of a McCormick line called Grill Mates, contains onion, black pepper, red pepper, sage, garlic, brown sugar, lemon peel, paprika, and "natural" flavor. Companies are taking aim at specific receptors now. McCormick's Bag 'n Season line targets the umami receptor (yeast extract, torula yeast, disodium inosinate, and disodium guanylate); the sweet receptor (sugar, molasses, corn syrup solids); and any number of the nose's four hundred aroma receptors (natural flavor).* It's not just McCormick. Crockery Gourmet makes a Seasoning Mix for Pork that is an even higher proof flavoring, containing hydrolyzed soy, corn, and wheat protein, sugar, as well as natural and artificial flavors.

---

* McCormick's line of jarred spices contains only dried herbs and spices.

Durkee Steak Dust contains natural flavor and an umami cluster bomb of hydrolyzed corn, soy and wheat protein, autolyzed yeast extract, and disodium inosinate and guanylate.

Flavor chemicals of one sort or another being added to, well, practically everything. Chocolate chips have "flavor" added. So do salad dressings. Ham, along with nearly every other deli meat and knackwurst sausages and turkey dogs, has flavor added. Bush's Baked Beans, which are made according to a "Secret Family Recipe," contain "natural flavor." (Grandpappy's gas chromatograph must have been out in the barn next to the chuck wagon.) You can even find curry sauce, an ingredient that has never appeared next to the word "bland," with added flavor. Raw sirloin burgers, frozen pizza, oatmeal, coconut cream pie, french fries, Earl Grey tea, herbal tea, imported pasta sauce from France, canned soup, Hamburger Helper, steak sauce, tzatziki sauce, breakfast cereal, chili, ready-to-bake raspberry turnovers, chipotle chunk white tuna in olive oil, crackers, yogurt, juices with no added sugar, ice cream, frozen burritos, dried cranberries, granola, and marinated raw salmon all feature synthetic flavor. All the things that used to taste good on their own aren't good enough on their own anymore.

Why is this happening? Why are the seasonings we put on our food becoming more and more like the ones we put on potato chips?

One popular answer claims we have finally thrown off the yoke of plain and boring victuals. We like food that thrills us, maybe even hurts a little. If our meat-and-potatoes grandparents had only known about dukkah, sriracha, and curry, they would have added a jar of each to their spice rack. Actually, they would have added a first-floor addition to make room for a bigger spice cabinet. It's called progress. Just like cars, phones, air travel, clothing, computers, and dishwashers, food is not only much better than it was fifty years ago, it's also way cheaper.

There is a lot to like about this theory. It is edifying to think of oneself as "cultured." Let's all laugh at the artless cuisine of our forebears. Who would willingly give up sriracha?

But it's wrong. For one thing, there was such a thing as curry fifty years ago. Curry is right there in 1902's *Ideal Cook Book*—curried chicken, curried eggs, curried steak, mulligatawny soup. Just like mace, sage, thyme, cloves, and every other herb and spice, curry powder was there if your great-grandmother wanted it. In 1918, Americans went through 310,000 pounds of mace. They liked spices. They used spices. But not every dish needed spices.

Julia Child, do not forget, declared that good chicken "should taste like chicken" and be "an absolute delight to eat as a perfectly plain, buttery roast, sauté, or grill." "Prodigal use of herbs," declared the Italian cookbook legend Marcella Hazan, is a "dead end." Garlic, she says, should be used "to advance flavors, not to club them senseless."

The real reason food companies keep jacking spices and taking aim at specific receptors is that we need them to. Chicken is different now. Tomatoes are different now. Corn, pork, wheat, strawberries, broccoli, collard greens, melons, barley, celery, lettuce, and radishes are different now. All food is different. It's been diluted. It's gotten worse. It needs help.

What else could explain the fact that the same industry that started adding flavor chemicals to margarine to make it taste more like butter is now adding flavor chemicals to actual butter? How did butter become the new margarine?

Flavor dilution is how. In 1948, the average dairy cow produced just over sixteen pounds of milk per day. It spent its days outside eating grass and dropping turd on the good green earth, and during winter it ate plenty of hay. Today, the average dairy cow lives the broiler chicken life, packed into a barn of airplane hangar proportions eating an unending supply of corn, soybeans, and roughage. The average dairy cow now produces more than seventy pounds of milk per day, and the very top-performing Holsteins crank out nearly two hundred, a udder-stretching 1,200 percent increase over the 1948 average. Modern milk is like modern chicken and modern tomatoes—bland and watery.

Diluted milk is one reason almost all strawberry yogurt contains "natural flavors" (and a whole lot of sugar). Another is that a piece of California land that grew a pound of strawberries a year back in 1948 now produces five and a half pounds. And while we're on the subject of California, since the mid-1940s, onion, nectarine, and table grape yields are up by almost 200 percent; celery and garlic are up by around 250 percent; artichokes, beets, broccoli, and cantaloupe by around 300 percent; and tomatoes and almonds by 500 percent.

A standard acre of American farmland is now producing more than three times as much rice than it did in 1948, four times as much corn, three times as much potatoes, two and a half times as much wheat, and twice as much soybeans. Hens lay twice as many eggs, pigs are 25 percent bigger yet 25 percent younger, and a beef cow is just half as old but produces 60 percent more meat. Decade after decade, we keep squeezing out more food from the same parcel of turf. We are paying for it in flavor.

There is a huge flavor problem in our agriculture. But that's why we have flavor solutions. We use 500 percent more herbs and spices than we did in 1918. We use "flavors." It takes so little "flavor" to flavor a drink, a granola bar, a serving of yogurt, or even a bag of Doritos, that the cost is often measured in fractions of pennies. An industry source I spoke to said it costs less than 5¢ to flavor twenty-four cans of Coca-Cola. Another said $12.50 buys you a pound of a "natural" yogurt flavoring— enough to make more than a thousand servings taste like something they are not.

Exactly how much flavor does the typical American eat? The government has no idea. The FDA doesn't track consumption, and neither does the USDA. The one organization I found that can put a number to it is a market research firm called Euromonitor International. According to their estimates, the American flavor market stands at 605 million pounds of flavoring every year. (This does not include MSG and the rest of the umami brigade, which stands at 190 million pounds.) That works out to two pounds of chemical deception for every man, woman, and child.

The industry that can flavor a serving of yogurt for 1¢ and a thousand cans of pop for $2 is worth $2 billion in the United States and $10 billion globally. *Allured's Flavor and Fragrance Materials* is 480 pages, the company Wilhelm Haarmann founded has an inventory of 50,000 flavorings, the world's biggest flavor company, Givaudan, has 20 flavor factories, and global flavor production is estimated to be more than 1.4 million tons.

This is the simplest and truest story of what has happened to food. It keeps getting blander. We keep adding more flavor. Year by year, the food we eat is becoming more and more like Doritos.

What happens to us when we eat it?

FOUR

Big People

J EAN NIDETCH had a word for foods so delicious they were especially hard to resist. She called them "Frankenstein" foods. They brought on a state of wanting so powerful she became an uncontrollable monster. Of all the foods Nidetch craved, none brought out her inner Frankenstein more than cookies. And the most terrible cookies of all were the chocolate-covered marshmallow cookies known as Mallomars. Nidetch loved Mallomars so much she would lock herself in the bathroom and feed on them in secret and then hide the empty boxes in the laundry hamper.

A half century after Nidetch's Mallomar binges, scientists had developed a technology that could see cravings erupting, like solar flares, inside the human brain. In early 2008, a research team at the Lewis Center for Neuroimaging at the University of Oregon measured just such a craving in a nineteen-year-old college student we will call Debbie. Debbie had her head inside a very large, very expensive round magnet called an MRI scanner when an image of a chocolate milk shake was flashed before her eyes for two seconds. As soon as Debbie saw it, certain parts of her brain became "activated," which is to say they drew in lots of blood as millions of neurons were fired. These regions—the left medial orbitofrontal cortex, anterior cingulate cortex, and three other small, curly pockets of gray matter—are all associated with "motivation." And the functional MRI (fMRI) showed them glowing a bright yellowy orange, like coals in a hot fire, indicating those parts of her brain were churning through quite a lot of blood. She was experiencing "incentive salience," the scientific term for a Frankenstein craving, or a heightened state of "wanting."

Debbie got what she wanted. Five seconds after the image of the milk shake flashed, actual chocolate milk shake was squirted into her mouth through a syringe pump. The cold burst of liquid chocolate lasted five seconds. Now her orbitofrontal cortex, which is associated with "reward," was glowing hot.

Despite the free milk shake, Debbie wasn't having a good day. The whole reason she'd been invited to do the brain scan, in fact, is that she enrolled in a program to help maintain a healthy weight. At five foot four and 160 pounds, she had a BMI (body mass index) of 27.5, which made her officially overweight. Before the scan, Debbie had responded to a series of statements about her eating habits that dredged up all sorts of unpleasant emotions:

- ▸ I find that when I start eating certain foods, I end up eating much more than planned.

- ▸ I find myself continuing to consume certain foods even though I am no longer hungry.

- ▸ I eat to the point where I feel physically ill.

- ▸ I spend a lot of time feeling sluggish or fatigued from overeating.

- ▸ There have been times when I consumed certain foods so often or in such large quantities that I spent time dealing with negative feelings from overeating instead of working, spending time with my family or friends, or engaging in other important or recreational activities I enjoy.

▶ I have had withdrawal symptoms such as agitation, anxiety, or other physical symptoms when I cut down or stopped eating certain foods.

▶ I have found that I have elevated desires or urges to consume certain foods when I cut down or stop eating them.

Response choices started at "never" (0 points) and went up to "4 or more times a week or daily" (4 points). Debbie's answers were consistently toward the high end. The questions are part of a standard survey called the Yale Food Addiction Scale, and, just as its name implies, it measures the degree to which people's eating habits resemble classic addictive behavior like smoking cigarettes, snorting cocaine, or shooting heroin. The higher a person's score, the more his or her eating behavior resembles substance abuse. Of the thirty-nine participants in Debbie's study, fifteen scored high on the scale. Debbie was one of them.

Not everyone believes food can be addictive the way drugs are. Drugs are highly potent neurotoxins, while food is just, well, food. That's why this particular study was carried out, to further demonstrate the way the brains of people who can't control their eating are like the brains of people who can't control their drug use. And Debbie's brain scan, along with the scans of her fellow food addicts, did just that.

To understand the surprising nature of Debbie's problem, it helps to know "Sarah," who also took part in the same fMRI study. Sarah is the same height as Debbie but weighed twenty pounds less. Her BMI of 23.8 was "normal," and she scored much lower on the Yale Food Addiction Scale. For Sarah, statements like "My food consumption has caused significant psychological problems such as depression, anxiety, self-loathing, or guilt" were more likely to be answered with "never" or "once a month."

You might expect that the differences between these two women asserted themselves when the milk shake was injected into their mouths. Debbie's reward center probably lit up like a raging volcano, whereas

Sarah's pleasure was contained. What else could explain why Debbie is always eating more than she planned, eating to the point of feeling physically ill, and eating despite knowing it's bad for her? The woman clearly loves food too much. She eats and eats and eats for the same reason junkies shoot heroin. It makes her feel good.

This prejudiced view, however common, is not the story told by Debbie's and Sarah's brains. It turned out that Debbie and her food-addict-type peers didn't derive any more pleasure from the milk shake than Sarah and her "normal" peers. The milk shake tasted just as good to everyone. The biggest difference between these two groups, in fact, had nothing to do with *drinking* the milk shake. It was all about the anticipation generated by the picture. The food-addictive types *wanted* it more. When the image of the milk shake appeared, Sarah got a bit excited, but Debbie got extremely excited. Her brain lit up like Las Vegas during a power surge. Debbie's problem isn't that she enjoys milk shakes too much. Her problem is that she wants milk shakes too much.

In other words, fat Debbie doesn't love eating food more than skinny Sarah. Fat Debbie craves food more than skinny Sarah. And then, when she finally eats it, it doesn't taste as good as she expected. It fails to live up to what she had anticipated. The milk shake isn't satisfying, so Debbie, hoping to enjoy an experience equal to her expectation, drinks more.

And with that, the study, titled "Neural Correlates of Food Addiction" and published in the *Archives of General Psychiatry*, proved its point. Classic addiction is also considered a disease of craving. Smokers crave cigarettes a lot more than they like smoking them just as alcoholics crave drinking more than they enjoy drinking and just as heroin addicts spend years chasing the transcendental peak of their first heroin high but never quite achieve it. So it was with Debbie and the others who scored high on the Yale Food Addiction Scale. Food never lives up to its billing. The required dose gets bigger and bigger. Their palates become altered. Junk food loses its ability to please them and healthier foods, as one of the study's authors, Ashley Gearhardt, put it to me over the phone, "don't have a chance." They get sucked into a ruinous vortex of uncontrollable eating. Incentive salience rages like a grease fire, and nothing can put it out. Their food-craving brains look alarmingly like the drug-craving brains of drug addicts.

FOOD-ADDICTED brains are not happy brains. We know this because in the fall of 2007, at the Scripps Research Institute in Jupiter, Florida, ten white rats lived the all-you-can-eat life and regretted it. The rats spent forty days enjoying a never-ending buffet of coconut pecan Pillsbury Creamy Supreme Frosting, Sara Lee Classic New York Style Cheesecake, Sara Lee All Butter Pound Cake, Nutella, Johnsonville Beddar with Cheddar Smoked Sausage Links, Peter Pan Creamy Peanut Butter, Microwave Hormel Bacon, and Hostess Ding Dongs. The rats did exactly what you'd expect. They ate. They *really* ate. By the end of the forty days, the rats were superbly corpulent, each gaining an average of around 165 grams, almost doubling their body weight.

These rats, however, were not indiscriminate gluttons. They were more like little whiskered connoisseurs. They quickly homed in on the foods they liked: cheesecake and bacon, which just happened to be the most energy-dense foods available. But the rats were particular about their bacon. They liked it cooked medium, so that none of the fat was rendered. They would nibble the streaky portions and avoid the crispy bits. They wouldn't so much as lick food they didn't like, leaving untouched portions on their food tray in what looked like a pointed message to the chef. Peanut butter was so universally unappetizing it had to be taken off the menu.

These rats had it good. They experienced none of the taxing stresses of ordinary rat life—no cats, no scarcity of food that required daily foraging expeditions. They lived in a dry, clean cage with lots of paper shavings, perfect for burying food (something rats love to do). Every day, a giant gloved hand entered each cage and removed the old, excrement-covered bedding and laid down clean bedding. The temperature was always a perfect seventy-two degrees. And there was endless free room service—an unceasing plentitude of some of the most deeply pleasurable victuals known to rat kind, all of it gobbled in blissful ignorance. There were no warnings from the surgeon general about the perils of obesity to spoil the joy of Ding Dongs, no anxiety about how they would look at the beach, no omnipresent barrage of celebrity rats with killer bodies to torpedo their self-worth. These rats enjoyed a life of supreme indulgence. Could it get any better?

It sure could. Despite all that food, these rats were not happy. They were desperate and unhappy in a way only addicts can be.

The scientists conducting the study didn't measure only body weight, they also measured the rats' state of mind. One way they did this was by giving rats an electric shock. Every time the shock was about to be delivered, a light would go on. Rats are smart enough that they eventually learn what the light means: a shock is coming. A normal rat will tense up and freeze to minimize its contact with the cage floor (where the shock is received). The food-addicted rats didn't do that. They were willing to suffer through an unpleasant shock if it meant they could keep on eating. This kind of compulsive behavior, as Paul M. Johnson, one of the study's authors points out, is one of the hallmarks of addiction.

Johnson and his adviser, Paul Kenny, a leading researcher in addiction, took another group of rats and fed them the same diet, only this time instead of subjecting them to electric shocks, they measured their overall state of happiness by applying electrodes to their brain (a technique called intracranial self-stimulation). And what they found is that the rats were very happy indeed, at the beginning of the experiment, at least. They loved all the cheesecake, bacon, cheesy smoked sausage, Nutella, frosting, Ding Dongs, and pound cake. It didn't last, however. Not only did their happiness wear off, it reversed itself. Eventually, the rats became chronically unhappy. After several weeks of indulgent eating, their food didn't make them feel good anymore. They were gorging on rich food to get back to "baseline." It just made them feel normal now. Eating simply dulled the pain of what had become an unpleasant existence. The rats were as miserable as they were fat.

SALT, SUGAR, FAT—perhaps you have already been introduced to the demons lurking in the cheesecake, bacon, cake frosting, and Ding Dongs. These are the evil, druglike substances that bring on the ruinous behavior driving the obesity epidemic.

At least, this is the latest take on the food problem. And what a refreshing take. After scrabbling around for the single nutritional villain that's killing us, or seeking salvation in grapefruit or sauerkraut, we have finally, as a culture, begun asking why everyone wants to eat so much.

Salt, sugar, and fat are what psychologists call reinforcers. They trigger bursts of the potent neurotransmitters and activate the same brain circuitry as heroin and cocaine. Sugar is the worst. We are hardwired to love sweet. Babies fresh from the womb smile when their sweet receptors are triggered by sugar. Salt and fat similarly activate brain regions associated with desire and reward—and not just in Debbie but in all of us. And when you mix them all together, look out. The food companies know this. The food companies, whose profit is directly related to the amount of food people eat, have been quietly amping up the amounts of salt, sugar, and fat in our foods, and the results are obvious.

Salt, sugar, and fat is a good answer. But it sure isn't the whole answer.

What about umami? There must be a reason Americans are being fed 190 million pounds of MSG, disodium inosinate, and other umami triggers every year. Food companies are not doing it in the vague hope that it might make people like their food more. They know it does something. But what?

There isn't as much brain-imaging work on umami as on other ingredients—the field is still young and these studies are extremely expensive—but there is some. So far, it has been observed that umami stimulates the orbitofrontal cortex, a brain region you will recall from Debbie's stint inside the fMRI, which is involved in matters of choice—dilemmas like "Do I want $5 today or $50 a week from now?" or "Do I want to eat another chip?" In obese people, the OFC lights up around delicious food, and when it becomes damaged, people become impulsive. Umami also activates the insula, a similarly important brain region when it comes to addiction. For example, if a smoker has a stroke and his insula is damaged, he won't be a smoker anymore.

There is also the matter of carbs. (Which is to say, carbohydrates other than sugar.*) They, too, are habit-forming. Just ask the sixty-one overweight-to-mildly-obese Chicago-area women who answered an ad offering $350 to "carbohydrate cravers" who wanted to participate in a research study. To qualify, these ladies had to experience afternoon or evening urges to eat "carbohydrate-rich, protein-poor snacks between meals at least four times per week" in what the study termed "emotional-eating episodes." Each woman visited a lab at the University of Illinois, where she was asked to recall a sad memory while simultaneously listening, through headphones, to "Russia Under the Mongolian Yoke" (a piece of music that is even more miserable than its name suggests). Afterward, the women were given one of two beverages: one high in carbs and the other higher in protein. Both beverages had been declared "equipalatable" (equally tasty) by an impartial panel of non-carb-craving adults.

The results were clear. The carb-craving overweight women loved the high-carb drink. It made them feel better. Their unhappiness "declined sharply" for two hours and began to rise again only at the 135-minute mark. Even three hours later, they still felt better than during "Russia Under the Mongolian Yoke." (The protein beverage didn't budge their mood.) Simply put, when these women were down, carbs gave them

---

*Though sugar is a carbohydrate, I will henceforth use the word "carbs," as it is in common parlance, to refer to carbohydrates other than sugar.

a lift. As the study's authors put it, the drink "displayed indications of abuse potential."

The important word here is "potential." Debbie, don't forget, weighed only 160 pounds. She was overweight, not extremely obese. She was showing signs of addictive behavior, but she wasn't anywhere near as far gone as those rats in Florida—not yet, anyway. This illustrates one of the subtler aspects of addiction that's generally lost on everyone but the people who research it. Substance abuse isn't a single extreme behavior. It's a continuum. Not everyone who tries addictive drugs like heroin or cocaine turns into a depraved fiend. About a third of people who try smoking become smokers, for example, and a quarter of people who try heroin become addicts. Many become "casual users." They experience aspects of addiction—cravings, regret—but drugs don't take control of their lives. Some drinkers, similarly, are just weekend bingers, and not everyone who struggles with food cravings is a morbidly obese food addict. Even the merely overweight wrestle wicked desires.

There is another difference between food and drugs. Eating isn't a lifestyle choice. When it comes to food, you can't just say no. The choice is starve or be a user. We used to be casual users. Now, nearly 7 percent of Americans have a BMI consistent with full addiction and millions more are using more than they ought to. It looks an awful lot like food has gotten too pleasurable. We can't resist. And it's all salt, sugar, carbs, umami, and fat's fault.

Well, it's actually worse than that. Because to place the blame only on those five ingredients is to miss a defining point about food and how we think about it.

Jean Nidetch didn't crave sugar. She didn't find herself stricken by abstract thoughts of umami. When the ice cream truck rolled down the block, did she think, "Calories in the form of sugar and mono-unsaturated and saturated fat"? The woman craved food. When food companies come to McCormick in search of what its customers want, McCormick doesn't sell them a calorie solution or a carbs solution—it sells them a flavor solution. Eating is a behavior driven by an expectation of pleasure. And the mental vocabulary of those desires is not salt, sugar, or any other class of nutrients. We crave flavors. Flavors are what make food seem like food.

Salt, sugar, and fat just don't seem like food all on their own. If I were to take some lard, roll it into balls, and dust the balls with sugar and salt, and then sell these prized morsels to hungry customers, I would not be lauded as a twenty-first-century food product development genius. Deliciousness is not that simple. And the reason is that my salty-sweet lard balls—even if I coated them in breadcrumbs and plopped them in the deep fryer—would be bland, and all the sugar, salt, and umami in the world wouldn't save them. What they lack is the same thing Wilhelm Haarmann's hometown is famous for: aroma. Add some "vanilla" or "cherry" to those salty-sweet lard balls, and then we're in business.

Salt, sugar, fat, carbs, and umami aren't new. All were widely available prior to the obesity epidemic. (Of the five, umami has probably experienced the greatest leap in availability.) Back before Arch West pulled into that taco shack in Southern California, human pleasure circuitry was regularly being hijacked by Fritos, potato chips, cookies, fritters, french fries, ice cream, fish sticks, hamburgers, and the like. There are addiction-worthy cakes and frostings as far back as 1902's *Ideal Cook Book*. And yet, extreme obesity was a fraction of its present rate.

It is tempting to view the increase in obesity as the result of an increase in dosage, that the big food companies have created the snack equivalent of crystal meth and gotten us all hooked. The truth is, in fact, more strange. The way to get more addicts isn't just to make an already potent drug even more potent. The epidemiology of addiction shows you can also get more addicts by making a powerful drug *more available*. Perhaps the best illustration of this is narcotic use during the Vietnam War, when an astonishing number of American soldiers were exposed to high-quality opium and heroin. In 1970 and 1971, almost half tried one and/or the other. Of that group, only a quarter were "experimenters." The rest used regularly and half reported feeling "strung out" and addicted while in Vietnam. Even more astonishingly, when their tours ended and they came back home, America was not overrun by a zombie army of heroin fiends. Only around 10 percent of Vietnam addicts remained addicts when they got back home. The others gave it up. They didn't join support groups or enter a methadone program. They just stopped. Their abusive behavior was, to a large degree, a matter of circumstance.

This is very much how the rise in obesity appears. People like Debbie—people with abuse potential—are being exposed to more abusive food. Seen through the lens of alcohol, it's not that we made whiskey stronger. It's that we secretly replaced everyone's beer with whiskey. Occasional users have become heavy users. And heavy users have become addicts.

The news is better than it sounds. If thousands of American GIs can kick heroin, then tens of millions of eaters can stop abusing food. For that to happen, however, we need to understand how food has changed. And to simply say there's more junk food now than there used to be is to miss the breadth of the way food has been altered. Yes, part of the problem is junk food. There's more of it, and it's more alluring than ever. But nonjunk food is a bigger problem. It isn't as flavorful as it used to be, which has the inverse effect of making junk food yet more enticing. Even worse, we're turning real food into junk food. Thanks to its off-putting insipidness, we coat it in calories, drench it in dressing, and dust it in synthetic flavor. The more bland it becomes, the harder we try to make it seem real. The first commercial gas chromatograph went on sale in 1955, and within a few decades its effects were rippling through the population.

Arch West's original Doritos, don't forget, featured three addictive substances: salt, fat, and carbs. They failed. They weren't saved by more salt, fat, and carbs. They weren't saved by sugar. They were saved by flavor. They were saved by a simulation of a taco.

THERE IS one weight-loss gimmick that seems conspicuously absent from the canon: The Livestock Diet. Perhaps a better name would be The Anti-Livestock Diet: How Not to Eat Like a Pig. Dieters could learn a lot from what pigs, cows, and chickens eat. Whereas humans want to stay fit and vigorous well into old age, for farmers the life of a chicken, cow, or pig boils down to the following: Get big as fast as possible and die young. The last thing a pork farmer wants is a barn full of trim, long-lived pigs that extol the virtues of prunes.

If we've spent all this time and money fruitlessly trying to lose weight, would we be better off figuring out how to gain weight? Perhaps understanding how pigs, cows, and chickens crank on the pounds can shed light on what humans are doing wrong.

One way to get a pig fat is to feed it fat. But it's not the only way. Carbs also make pigs fat. It comes down to calories. And that's what pigs get. Calories. They're usually in the form of carbs, which is to say corn, millet, barley, and other starchy grains, because carbs are cheaper than fat, fat is hard to store (it goes rancid), and it generally makes a mess of the barn. Pigs also get some protein—soybeans are popular these days—and a little fat and a sprinkling of vitamins and other "essential nutrients." Their diet is a lot like the "high-energy diet" broiler chickens started eating in the 1950s, so similar that farmers often give pig feed to their chickens and vice versa.

The industry refers to this kind of feed as "concentrates." Think of it as a big intense wallop of calories and protein. Just like a fast car needs high-octane gasoline, a fast-growing animal needs high-octane feed. And this feed is so good at getting animals big quickly, in fact, that if cattle, pigs, and chickens aren't killed young, the same thing happens to them as to humans who eat too much. They become obese, and then metabolic disease sets in. Chickens get so heavy they become knock-kneed or bowlegged, and sometimes they gain weight at such an astonishing clip they suffer from pulmonary hypertension syndrome, in which the blood vessels and lungs can't keep up with their heart rate and their abdomens fill with a strange yellow fluid and they die. Cattle can keel over from stomach bloat, acidosis, or liver failure. Of the three, pigs seem to be best suited to packing on the pounds, but calories eventually catch up with them, too.

So there you have it. The lesson of the Livestock Diet: Lay off the concentrates. Keep away from the human feed trough filled with high-calorie food.

Most people, of course, can't do that. And there's one other impor-tant lesson from the Livestock Diet that helps explain why so many hu-mans keep sidling up to the feed trough: palatants.

You can lead a pig or cow to a trough full of concentrates, but you can't always make it eat. That is a problem for farmers. It's an especially big problem when it's young cows and pigs that are being picky eaters, because in a worst-case situation, they can get sick and die. And even if they don't die, skinny piglets and calves have a tendency to grow up into skinny pigs and cattle. For farmers, pounds equal money And lost poundage is like lost income—you never get it back.

There are tricks to get around this. Jell-O powder is a good one. Add it to a bottle of milk replacer—baby formula for cows, basically—and a stubborn, sick calf that misses its mother will happily suckle on a plas-tic teat. (Strawberry-banana is said to be a favorite.) The art of getting livestock to eat more, however, has matured into a science. Farmers can go out and buy little packages of precisely formulated powder called palatant. In nature, pigs are weaned at three to four months of age. On a modern pig farm, they're weaned as early as ten days, which can lead to "weaning problems"—diarrhea, dehydration, pneumonia. It's a seri-ous problem. As an industry newsletter puts it, "The main challenge of the pig industry is to promote feed intake of the piglet at weaning by all the possible means." There is a flavor solution. Palatant is added to milk replacer, or mixed into a trough filled with corn, soy bran meal, whey powder, porcine plasma protein, blood meal, and fat. Watch the critters dive in. Purina's UltraCare piglet feed "tastes so good they won't be able to keep their noses out of it." Pancosma's TakTik X-IN is a "state-of-the-art synthesis of sweetening and flavouring taste enhancement" designed to "increase feed intake." QualiTech's line of Feedbuds flavor systems comes in twenty-five different flavors, including Butterscotch Savour, Caramel Delight, Very Berry, and Anise-Fenugreek.

If palatants remind you of the flavorings produced by flavor factories, that's because they are the same thing. Palatants are created by flavorists in labs who combine aromas with sweeteners while channeling their inner piglet or calf.

Drew Vermeire is an animal nutritionist based in Lake St. Louis, Missouri, who uses palatants seven days a week to help farmers get their calves to grow. The blander the feed, he says, the more effective the palatant. Typically, they increase "intake"—which is to say eating—by around 5 percent. (A study out of South Africa, which Vermeire says is not unreasonable, saw an improved weight gain of 30 percent in lambs fed palatant.) That might not sound like much, but over weeks and months, it adds up to serious poundage. Vermeire estimates at least 75 percent of cattle eat palatant at some point in their lives. (Two swine nutritionists I spoke to said the figure is the same for pigs.) So do a lot of dogs and cats, incidentally, whose food is also laced with palatant. Several years ago, a company that manufactures cat palatant gave Vermeire a sample, a tan-colored powder. He dabbed some on a small piece of tissue and presented it to his daughter's cat, Sassy. "The cat inhaled it like it was a dead mouse," he says, "and then sat there licking its paws."

THIRTY-ONE PERCENT. That, you will recall, is the proportion of Americans who are not overweight or obese or extremely obese. Less than a third of Americans are, as Jean Nidetch put it, "slender." Just a few pages back, the number seemed depressingly low. Now it seems high.

How do they do it? How do these people manage to resist all that sugar, fat, salt, carbs, and umami? No one else seems to be able to.

So let's add one more villain to the list of obesity-inducing food intoxicants: flavorings. These apparently harmless, noncaloric chemicals look an awful lot like a missing piece of the obesity puzzle. Take the milk shake Debbie and Sarah drank. It wasn't just a gelid mixture of sugar and fat. It contained Hershey's Chocolate Syrup, which is laced with human palatant: vanillin and artificial flavor. Jean Nidetch's Mallomars contain artificial flavor. Of the eight foods the rats in Florida gorged on, six contained natural or artificial flavor. Of the two foods that didn't, the rats wouldn't touch the peanut butter. And that carbohydrate-rich beverage those carb-craving Chicago-area women so enjoyed? The one that lightened their mood for an entire afternoon? The women didn't know it at the time, but they were wearing flavor goggles. Without human-made flavor, it would have been a bland, pasty, vaguely sweet, and very watery beverage. Thanks to "natural" and artificial flavor, it tasted like fruit punch.

None of this should be in any way surprising. As instincts go, eating is number one. Is it really so shocking that an animal with a big brain would use that brain to spike deliciousness? The Dorito Effect was inevitable. It took ten thousand years, but we eventually nailed "taco." We finally figured out how to make ourselves wondrously fat. It was just a matter of technology. Farmers had been chasing the broiler dream for decades. As far back as 1894, one W. G. Anthony of Smyrna, Delaware, did a booming business force-feeding chickens with a cramming machine until their throats swelled up to the size of an orange—the only reason his business collapsed is that the price of skim milk went up. In 1948, poultry technology finally caught up to human ambition. If cats were the ones with big brains and furry little opposable thumbs, there would be factory farms raising plump, fast-growing mice of tomorrow whose bland flesh would be dusted in palatant in mouse-processing factories. If cats could eat Doritos, they would. And they'd be just as fat as we are.

We're done for. The rise in obesity is the predictable result of the rise in manufactured deliciousness. Everything we add to food just makes us want it more. And no matter how hard we try, we can't make our outsized desires go away. If anything, we're lucky, inexplicably so, that only 8.3 percent of women and 4.4 percent of men have a BMI consistent with total food addiction. But remember the children. They're getting fatter, and just like with calves, piglets, and rats, a husky-size youth grows up to be an XXL adult. The percentage of slender Americans will gradually work its way down to zero. One day, there will be no more Sarahs, only Debbies, and everyone will walk around with an insulin pump. If you haven't yet bought stock in the companies that make Lipitor and sweatpants, now would be a good time.

IT DOESN'T have to be this way. We could live in a world where food tastes very good and the people who eat it are not fat. But nothing is going to change until we get beyond a fundamental mistake in our thinking about food. Pleasure may help explain why people are eating too much. But the puritanical notion that deliciousness is sinful and that we can solve the food problem only by extinguishing the pleasure we receive from food is doomed to fail. Synthetic flavor might be the salesman in the fancy suit that sells our brains the calorie-rich fat and carbs we're eating so much of, but real flavor—the authentic version produced in nature—is our only road to salvation. I will say it again: Real flavor is our salvation. I offer you these two pieces of evidence:

1. **The human nose.** The best aroma-sensing equipment money can buy—an Agilent 7890 gas chromatograph paired with a LECO Pegasus HRT mass spectrometer—will run you $350,000. It can take that equipment hours to "taste" a single substance, and it can't even say if it tastes good or not. It can tell you only what's in there. And even then it will still miss stuff.

The human nose is instant. The human nose is technology money can't buy. And the human nose, contrary to popular opinion, was not designed to sniff. A dog can sniff eight times a second. A dog, like most mammals, has a long snout that cleans, warms, and moisturizes incoming air, which improves sniffing. The human nose does none of these things. A human can't close its eyes and tell where a deer is just by smelling the air, or if there's a tiger crouching in the bushes.

But a human can close its eyes and bask in the character of the food in its mouth in a way a dog cannot. Our smell apparatus appears to be designed for maximal food appreciation, as Yale neurobiologist Gordon Shepherd points out in his excellent book *Neurogastronomy.* Our cavernous nasal cavity, with its ten million smell receptors, is like a sensory echo chamber. Whereas a dog's aroma sensors are pointed at incoming air, ours are perched at the top of the cavity, situated, it seems, to sense both incoming and outgoing air. Those aroma sensors send signals to the brain—the big, complex, multitalented human brain—where they engage continents of gray matter. No activity stimulates the human brain to the degree that tasting food does. A human could never walk through a tropical forest and locate a dark, luscious vanilla bean by twitching her nose. But only a human could taste a sample of imitation extract in Hunt Valley, Maryland, in the late 1970s, exhale through her nose, and feel secure, satisfied, pleasant, and nostalgic and then announce, "I'm not getting the resinous note."

A question: Why did evolution endow us with a brain-hogging $350,000 flavor sensor and stick it right in the middle of everyone's face? Every piece of the human body is there for a reason. If genetics is about tradeoffs, it's obvious why evolution deemed feet, eyes, kidneys, and white blood cells worth keeping. So what's the nose there for?

2. **Treviso.** This small Italian city, which is bisected by three rivers and known for its beautiful stone bridge, has a curious food habit. It prizes a purple lettuce that is barely caloric but so wincingly bitter that in other parts of the world is quietly ushered to the edge of the plate: radicchio. In the Veneto region, radicchio is protected the way champagne is protected in France—only a handful of cities and villages are deemed capable of producing the genuine article. I once met a native of Treviso and, naturally, asked her if she liked radicchio. Her answer, and I am quoting directly, was, "I love it. I love it. I love it." The very best radicchio is *tardivo*, which is picked in late fall and considered so delectable the city holds an annual radicchio party where people eat radicchio frittatas, risotto, crostini, and even radicchio salad, which contains the notoriously bitter raw leaves, and the elderly can be heard complaining that modern radicchio is not as bitter as it used to be.

Italians take food seriously. People from all over the world endure transoceanic flights and jet lag just so they may eat as the Italians do. So here is a surprising fact about the food-loving Italians: They are not very fat. Only around 10 percent of Italian adults are obese, compared to 35 percent of American adults. Despite the exalted pleasure eating brings, Italians somehow keep their desire in check. What gives?

The human nose. Slim Italians and their bitter radicchio. Obesity. All are linked by a question so simple it almost sounds silly. And yet, during our decades-long conversation about nutrients, no one has ever asked it. Its answer will forever change the way you think about food. This is the question: Why do things have flavor in the first place?

The foremost authority is a goat in Utah.

PART TWO

IF FOOD COULD TALK

FIVE

The Wisdom of Flavor

I T ALL began with the mysterious case of the delicious urine.

In the late fall of 1976, a twenty-five-year-old PhD student at the University of Utah named Frederick Provenza was beginning the first of three winters living in a trailer on a high-desert plateau called Cactus Flat. Provenza was there to see what would happen if you let ninety angora goats spend several months eating nothing but blackbrush, a dark, bushy shrub that grows in Utah and Arizona. The experiment was more interesting to Provenza than it was to the goats, who don't consider blackbrush much of a meal, and it wasn't even that interesting to Provenza—not in the beginning, at least—who was an outdoorsman from Colorado who grew up hiking and skiing and imagined he'd spend his career researching animals in the wild, not goats in the middle of nowhere. But there was research money in blackbrush. If goats spent the winter pruning back the old, woody twigs, new twigs, it was hoped, would sprout that were more tender and nutritious, creating better foraging land for cattle.

For company, Provenza had his wife, Sue (they met ski racing), the ninety goats, and wood rats, rodents that lived in great unkempt nests at the base of juniper trees. Wood rats are nocturnal, so Provenza didn't see them much, but he'd find stashes of their food in the engine compartment of his truck, and some mornings he'd discover a drowned one floating in the water trough.

One morning, Provenza was walking through the goat pastures when he noticed a bunch of goats crowded around the base of a juniper tree eating something. He walked over and found that the goats had "totally annihilated" a wood rat house. Big pieces of juniper bark, which the wood rats used like aluminum siding, were scattered everywhere, as if a tornado had blown through. Even stranger, the goats were gobbling up the soft nesting material—twigs, leaves—underneath. Provenza had never seen or heard of anything like this before. So he crouched down to have a better look, and that's when he smelled it: urine. Urine is many things—odorous, disgusting, occasionally hilarious, and a good source of nitrogen. And it was the nitrogen that got Provenza thinking, because there are microbes in a goat's stomach that can turn nitrogen into protein—just the thing a skinny goat subsisting on a poor diet of blackbrush needs to get through the winter.

The mystery didn't end there. There were six pastures in all, each with fifteen goats, but it was only in this particular pasture that the goats were hoovering up wood rat houses. It was as though they developed a taste for it. Their white muzzles would become dark with wood rat excrement, and by the end of the winter there wasn't a single wood rat house left in the entire pasture. Was it possible, Provenza wondered, that these goats were eating urine-soaked wood rat nests to help them digest blackbrush? Or did they just like the way it tasted?

The answer, it seemed, was both. At the end of winter, Provenza weighed all the goats. Every single one lost weight—that's how lacking in nutrients blackbrush is. But the group that ate wood rat houses lost the least, and they were in the best shape—healthier and more vigorous. It was as though they had some built-in ability to seek out the foods they needed. To Provenza, it appeared that these goats possessed "nutritional wisdom."

This was a radical notion. At the time, the only people who believed in flaky concepts like "nutritional wisdom" or that goats could be "in touch" with their inner needs were assumed to have been smoking something. Back on campus, where the stark and unbending rules of science reigned supreme, "nutritional wisdom" was laughed at. Goats, everyone knew, were born with a built-in palate. There was a list of things goats were programmed to like—alfalfa, ryegrass, oak brush, etc.—and near the bottom of that list was blackbrush. By bumbling their way through the world and eating the stuff they liked, goats managed to meet their nutritional needs. But there certainly wasn't intelligence at work. Goats ate for the same reason they breathed. Some primitive impulse told them to.

Goats were stupid. Goats had no idea what was good for them. For example, another time during his stint at Cactus Flat, Provenza came upon the graying skeleton of a juniper tree that had been struck by lightning. All around it young blackbrush bushes, whose shoots are red, were growing. The goats will just love this, Provenza thought, because young shoots are more tender and nutritious than old woody ones. But when Provenza unleashed the goats on this tasty treat, they wanted no part of it. They sniffed it, one took a bite, and then they walked. So much for nutritional wisdom.

Back at Utah State, Provenza described the incident to a senior faculty member, who laughed at the idiocy of goats and declared it was yet more proof that there was no such thing as nutritional wisdom. But the episode ate at Provenza. Over the subsequent two winters, he noticed that goats never ate new blackbrush shoots. The behavior didn't appear stupid so much as intentional. It looked like the goats were avoiding it.

A few years later, Provenza went to a meeting in Boston, where he met a chemical ecologist from the University of Alaska, a legendary figure named John Bryant who had a thick beard and had spent years in the backwoods of the North, and who was more open-minded about the inner lives of goats than the senior faculty back at Utah State. Provenza had a theory that the blackbrush shrub was producing a toxin to ward off animals from eating its tender new twigs. It made sense in the larger ecological picture. If the new shoots wanted to survive long enough to become mature twigs, then it was to the blackbrush's benefit to discourage goats and other critters from eating them. Perhaps their red color even served as a warning. Bryant agreed and offered to help get funding to do an experiment.

By January 1989, the funding was in place, the goats were hungry, and Provenza had a pickup truck full of young blackbrush shoots. The plan was to make a liquid extract of new blackbrush shoots, separate the extract into its constituent parts, and then apply each of those parts, one at a time, to the goats' feed. At an outdoor lab not far from campus called the Green Canyon Ecology Center, which is perched on the edge of the Cache Valley above the city of Logan, Provenza would feed the goats the different extracts one at a time. He confidently predicted that on one of those days, the goats would sense the toxin in their feed, refuse to eat, and thus prove that they possessed nutritional wisdom.

That day never came. As Provenza made his way farther and farther down the list of blackbrush extractions, the goats kept on merrily eating. Finally, on a cold and foggy winter morning, Provenza was down to the last extract, a tannin. (Tannins are a family of chemicals found in plants and are abundant in red wine.) According to his theory, the tannin was the toxin the goats were avoiding. It had to be—there was nothing else. But when Provenza poured the toxic breakfast into the goats' feed trough, the goats lowered their noses and started gobbling. Either the goats were eating poison or, back at Cactus Flat, they had been avoiding food that was perfectly healthy. No matter how you sliced it, two things were certain: Goats didn't know what was good for them, and neither did Provenza. Now there would be no paper to publish. All that prestigious grant money from the National Science Foundation was, at that moment, going poof. Outside, it was a wall of fog. Provenza couldn't see the city of Logan spread out below, or the sharp peaks on the other side of the valley. All he could see were the goats, twenty of them, reclined in the state of fine contentment that follows a good meal.

AT ALMOST exactly the same time Fred Provenza began asking his goat question, two scientists on the other side of the country were making a similar inquiry of wasps. Their question was not how a wasp decides what to eat. They already knew that, because these were parasitic wasps. Each one hatched from an egg and spent most of its life as a wormlike maggot eating a caterpillar from the inside out, until it metamorphosed into a wasp and mated. If the wasp was female, it had to go find a new caterpillar and begin the whole cycle afresh.

And that was the question: How did a wasp zooming over a field of tall greenery find an inch-long caterpillar hiding inside a fruit or clinging to the underside of a leaf? It couldn't hope to spot such a caterpillar from a distance, and it didn't have time to check every plant. A helicopter hovering over Malibu searching for celebrities faced better odds.

In the early 1980s, this was an important question. Parasitic wasps were being raised by the thousands in wasp factories, loaded into paper bags, and dropped out of airplanes over fields of cotton that had become overrun with caterpillars. The hope was that the parasitic wasps would wipe out caterpillar pests, thus eliminating the need for toxic pesticides. It sounded great in theory, but it wasn't working. When the paper bags hit the ground, the wasps didn't do what they were supposed to.

An entomologist named Joe Lewis and an organic chemist named Jim Tumlinson took a crack at figuring out what was going on. They thought wasps found caterpillars the way a bloodhound finds an escaped convict: by following its trail of scent. To test their theory, they put a wasp in a cage near a caterpillar and watched to see what would happen. If the encounter could be likened to a first date, this was a spectacularly bad one. The wasp got up and flew around. It landed on a light. Then it just sat there. And that was it. They did it over and over again with lots of wasps and lots of caterpillars, but the result never changed much. The caterpillar wasn't giving off so much as the gentlest suggestion of scent. When Lewis and Tumlinson considered the situation, it made sense. If wasps were good at smelling, then it would be in the best interest of the caterpillar, from an evolutionary point of view, to be scentless.

Next they put a wasp in a cage with a caterpillar sitting on a plant. The second date was even worse. If anything, the wasp was mildly interested in the plant. It might fly over to the plant, hover near it, perhaps even land on it. But it had no time for the caterpillar. It was anything but a convincing performance.

But here's what was convincing: a partly eaten plant. If the caterpillar munched the leaves of the plant it was sitting on, the wasp perked up. The wasp fluttered its antennae and took flight, zigzagging back and forth across the vapor trail before zeroing in on the caterpillar like it had radar. Even more strange, wasps were attracted to half-eaten plants with no caterpillar. It was almost as though the plant was talking to the wasp.

And that's exactly what the plant was doing. It was rudimentary and nonconscious, but it was communication. The way they do it, Lewis and Tumlinson discovered, was through smell. "The plant," as Joe Lewis puts it, "was squealing on the caterpillar that was eating it." The wasp listened.

A few hours after being attacked, a plant sends out a burst of chemicals that tell wasps it's being eaten by a caterpillar. This scent is so specific wasps can tell what kind of caterpillar is doing the attacking. One chemical burst might indicate a tobacco budworm, and another might indicate a corn earworm. To a wasp that laid its eggs inside corn earworms but not tobacco budworms, this was important news.

Plants possessed another, very different, chemical alarm, Lewis and Tumlinson found. This one says, "I'm wounded!" and it was nearly instant. This alarm warned neighboring plants that trouble was afoot, and the neighboring plants would begin stocking up their own arsenals of alarm chemicals. This alarm told the wasp exactly where the caterpillar damage was happening, which is helpful to a wasp searching for a tiny caterpillar on a four-foot cotton plant amid an entire field of four-foot cotton plants. Drawn in by the wafting caterpillar alarm, the wasp would pick up the wound signal, zero in, mount the caterpillar, and insert its stinger.

As for those billions of wasps being dropped out of airplanes that couldn't find caterpillars, they had many problems. Back at the wasp factory, the wasps were being hatched inside caterpillars that were fed pinto beans, corn, and soy meal (the caterpillar version of the high-energy diet), not cotton leaves, like out in farmers' fields. It turned out that when a wasp maggot eats its way through a caterpillar, chemical traces of the caterpillar's food imprint on the maggot and tell it what kind of chemicals to look for when it grows up and becomes a wasp. The wasps being hatched in factories and dropped out of planes were looking for the wrong chemicals. They hadn't been programmed to find caterpillars that were eating cotton.

And even if the odd wasp somehow figured out what a cotton leaf smelled like, there was another problem—one that, by now, should be all too familiar. Jim Tumlinson took four highly productive varieties of cotton and measured the strength of the chemical wound signals they gave off. Then he performed the same measurement on a wild cotton plant found growing in the Florida Everglades. The difference was striking. The wild cotton produced a chemical plume that was ten times stronger. Parasitic wasps were having trouble finding their hosts for the same reason tomatoes taste like cardboard: dilution.

And what, exactly, were those wound chemicals? Lewis and Tumlinson—who would both go on to win agricultural science's most prestigious honor, the Wolf Prize—measured that, too. They slashed a leaf with a razor blade, collected the wound odors, and analyzed them using gas chromatography. And do you know what one of the major chemicals they found was? Cis-3-hexenol. Humans and wasps, it turns out, both love the smell of cut grass. To a human, it's an integral part of a strawberry. To a wasp, it's the smell a plant gives off that says the caterpillar is over here.

Flavor is information.

OVER AT the Green Canyon Ecology Center, Fred Provenza's goats did not have an information problem. Their flavor system, as he would come to understand, was running perfectly. The problem the goats had, funnily enough, is the same problem Provenza himself had—a lack of information.

The morning of the goat disaster, Provenza paced back and forth inside a frigid cinder-block garage next to the goat pens as he and his research assistant weighed the options. It had taken months to purify the tannins, and at the rate the goats were inhaling supposedly toxic feed, they had only enough left to run the experiment one more time. And that's what they decided to do—feed the goats more of the tannin-laden feed the very next day. Looking back, Provenza isn't sure why they came to that decision. What difference was one night going to make?

A big difference. The next morning, Provenza filled their feed trough with more of that supposedly toxic feed, but this time the goats took one sniff and wouldn't go near it. The food they'd enjoyed just a day earlier was a no-go. The reason was toxicity. Every goat had spent a long, dark day struggling with nausea and malaise. When they smelled that tannin the next day, it came with a warning: This will make you feel sick.

Nutritional wisdom, Provenza discovered, was real. And just like actual wisdom, it's not something goats are born with. It must be learned.

PLANT SECONDARY COMPOUND. Remember these three words. Blackbrush tannin is a plant secondary compound. So is cis-3-hexenol. And so are vanillin and cinnamic aldehyde.

There are 45,000 plant secondary compounds that have been identified, give or take, and the grand total may be as high as a million. If you want to understand why "things" have "flavor," and why blurring that line has big consequences, it is worth spending some time to get to know these misunderstood and surprising chemicals. The term is a mouthful, but it's important. And ironic, because the most important of those three words is "secondary."

For most of the last two hundred years, scientists didn't know what to make of all the chemicals produced by plants. There were so many— more, it seemed, than a plant actually needed. Some chemicals, clearly, were indispensable. Cellulose, for example, gives plants structure, and chlorophyll lets plants absorb energy from sunlight. Without any one of these important chemicals, the plant was a sure goner. These were *primary* compounds. And there weren't many of them.

But what about the thousands of other chemicals that plants produced but didn't use? For a very long time, no one particularly cared. Ever since a German pharmacist isolated morphine from opium poppies in 1806, scientists were mainly concerned with one question: What other useful things do plants make? (Vanillin was just one among many answers.) After more than a century—a very successful century—of finding chemicals in plants, scientists began to ask why plants made all these strange chemicals.

Some biologists thought these chemicals were the accidental byproducts of primary compounds—the plant-as-fumbling-chemist theory. Others speculated these substances were some kind of waste product the plant couldn't get rid of—the plant-as-constipated theory. No one had much of a clue, and it hardly mattered. They didn't help a plant grow, or stay alive, or reproduce, which is the whole point of life. They were *secondary*.

In the 1950s, a Jewish refugee from Nazi Germany named Gottfried Fraenkel began to challenge the prevailing arrogance. Since all green plants were pretty similar as far as nutritional makeup, he wondered why insects were so particular about the leaves they munched. He noted, for example, that insects usually avoid plants in the Cruciferae family because they produce poisonous compounds known as mustard oil glycosides. But strangely, the European cabbage worm eats almost nothing but Cruciferae that produce mustard oil glycosides. There are, however, a few non-Cruciferae plants in the cabbage worm's diet, and, curiously, these also produce mustard oil glycosides. In fact, you could fool a European cabbage worm into eating any old leaf just by spreading mustard oil glycosides on it.

Fraenkel believed plants produced these strange chemicals for a very good reason: to ward off hungry insects. Most of the time, it worked. But some insects evolved a tolerance. And these insects, he postulated in the prestigious journal *Science*, not only tolerated the poisons, they developed a preference for them. A chemical message that says, "Do not eat me!" to some insects says just the opposite, "Eat me!" to others. And these lucky insects got the plant all to themselves. Fraenkel called these chemicals "trigger substances."

The paper, now considered landmark, was met by "icy silence," as Fraenkel would later recall. (And the silence had been going on for decades, because Fraenkel himself was drawing on the work of a long-dead colleague named Eduard Verschaffelt, who'd been ignored since 1910.) By the early 1970s, though, as outdoors types like Fred Provenza started getting PhDs, the thinking began to change. It was recognized, for example, that plants sure did produce an awful lot of poisons. Creeping indigo, to use just one example, contained a chemical called indospicine that causes liver damage, weight loss, and death in rabbits. If rats ate cycad nuts, a chemical called cycasin caused them to develop tumors, and a higher dose could finish them off in less than a week. Plants weren't bad at chemistry and they weren't constipated. Evolution had designed plants to be brilliant chemical strategists.

GOATS, TOO, were strategists. They had to be. If plants had evolved chemical strategies, any animal that went around eating them needed counterstrategies. No one wants to get poisoned by dinner, after all. So they evolved a system that let them figure out what was good for them and what to stay away from: nutritional wisdom. It explained why goats wouldn't eat new blackbrush shoots, and why, when they were nearly starving, they developed a taste for wood rat urine.

But how nutritionally wise is a goat? How well can its flavor-information system cater to its inner needs? Fred Provenza spent his career asking that question, and the answer he arrived at is: amazingly, exceedingly, and extraordinarily wise. Nutritional wisdom isn't a simple, rudimentary, and practical knowledge—eat this, don't eat that. It's more like a kind of genius.

In one typical experiment, Provenza looked at the mineral phosphorus, which the body uses to make things like bones and teeth and for a variety of functions (such as intracellular communication). Phosphorus is essential for life. Without it, you're done for. Scientists had long suspected that animals get phosphorus by liking salt. Salt itself is only one mineral—sodium chloride (also essential)—but in nature salt often shows up alongside other minerals. By liking salt, the thinking goes, an organism gets all its minerals. But Provenza didn't think it was so simple. He'd noticed, for example, that phosphorus-deficient cattle often lick and chew on old bones. There is phosphorus in old bones, but no salt. So how could liking salt explain bone chewing? It couldn't.

Provenza placed ten sheep in a pen and put them on a diet that was low in phosphorus. When the sheep developed a deficiency, he offered them some feed laced with maple flavoring. And then came the important part: After the sheep ate the maple-flavored feed, Provenza pumped phosphorus directly into their stomachs through a tube that went down their throat. This sounds bizarre, perhaps even cruel, but Provenza had to ensure the sheep didn't taste the phosphorus. He fed them maple-flavored feed and pumped phosphorus into their stomachs for six days.

He wanted to set up an "association" between the flavor of maple and the nutritional payload of phosphorus. A few days later, when these same phosphorus-deficient sheep were offered maple-flavored feed, they gobbled it up, even though there wasn't so much as a speck of phosphorus in it. To their bodies, maple flavor meant one thing: phosphorus. The less phosphorus in their blood, the more they liked the maple feed.

How do we know sheep aren't just natural-born maple lovers? Because Provenza had another pen full of sheep, and in that pen phosphorus was paired with coconut flavoring, not maple. And when these sheep became deficient in phosphorus, they went for coconut. In fact, when all the phosphorus-deficient sheep were given a choice between coconut and maple, they always went for the flavor they associated with phosphorus, which demonstrated the sheep weren't just looking for anything other than their phosphorus-deficient feed. When the phosphorus deficiency went away, so did the flavor preference. But the sheep didn't forget it. All Provenza had to do was make them deficient again and back came those very specific flavor cravings.

Provenza was measuring a phenomenon known as "post-ingestive feedback," which is a technical way of saying flavor preferences are learned (as with the goats and the blackbrush tannin). The sheep didn't like coconut or maple the first time they tried it. The preference set in once their bodies connected the dots between the flavor of what they ate and the needed mineral their bodies received.

Provenza did a similar experiment with calcium, and the same thing happened. And it wasn't just minerals. He got the same result with carbs. When lambs on a low-calorie diet tasted a flavor paired with a stomach drenching of carbs, they quickly learned to love that flavor. (Just like those carb-craving women in Chicago.) In another study, Provenza showed that it happened with protein, too. Sheep, in fact, were very particular about just how much protein they wanted. Growing lambs, which require more protein than mature sheep, formed a preference for more protein. Sheep, like goats, were woolly little nutritionists.

If animals were so smart, could they outnutrition a big-brained human with a PhD in animal nutrition? Provenza tested that, too, this time with calves. He offered calves in one pen a premixed feed that had been designed by a ruminant nutritionist to perfectly meet their nutritional needs. In the other pen, he put out all the foods that went into that high-precision mix—alfalfa hay, silage (fermented corn plants), rolled barley, and rolled corn—and let the calves figure it out for themselves. The winner? The "free-choice" calves. By the end of the experiment, they had gained more weight—these were growing calves, so growth was a desired outcome. The free-choice calves did a better job of filling their daily requirements for protein, energy, and nutrients than the human with the fancy degree.

IN 1998, Provenza's love of the outdoors took him on a six-week rafting trip down the Grand Canyon. Before departing, a ranger with the National Park Service gave a talk about safety and showed a video on the importance of drinking water. "You can't drink enough water," the ranger said, a comment that struck Provenza as "insane." The body, he thought to himself, is going to tell you what it needs. When Provenza went back a few years later to do it again, this time the ranger had a different message. Drink when you're thirsty. Some of the paddlers, it turned out, had followed the earlier advice so explicitly they had come down with a syndrome called water intoxication and had to be airlifted off the river and taken to a hospital. They had drunk so much water it became toxic.

Provenza began reflecting on the curious nature of toxicity. Plant secondary compounds tended to get a bad rap a lot of the time because they were so often poisonous. Provenza didn't think it was that simple. Out on the range, he noticed that goats and sheep would nibble all sorts of plants, many of which were mildly toxic.

This was not some casual observation. He recorded it down to the level of grams. In perhaps his most peculiar experiment, Provenza cut an opening in the throats of six goats—known as a fistula—and tied bags around their necks. (Provenza used anesthetic, followed humane handling guidelines, and ensured the animals were well cared for following the experiment.) As the goats browsed and swallowed, their chewed-up food fell into the bag, and at the end of the day Provenza rooted through the gooey mess to see what the goats were eating. (When he wasn't collecting, he plugged the hole and the goats ate and swallowed as usual.) The goats were eating a lot, fifty different kinds of plants in a single day.

Around the same time, Provenza began learning about the traditional—which is to say, peculiar—grazing practices of shepherds in France. The shepherds walked their flocks on ornate grazing circuits. The sheep would spend part of the morning nibbling these plants over here, but not before eating a bit of other plants over there. Certain combinations of plants, the shepherds found, worked best. They even described different "courses" to a meal: appetizer, main course, and dessert.

Provenza began to wonder, Was food in nature, the product of millions of years of evolution, really as binary as "good" and "bad"? If too much water could be bad, perhaps a little bit of something bad could be good.

Way back in 1959, Gottfried Fraenkel as much as said so by pointing out that certain insects were attracted to the poisonous compounds that frightened all the other insects off. The best-known example is monarch butterflies, whose larvae feed on milkweed plants that contain a toxic alkaloid (a type of secondary compound). The alkaloid builds up in the monarch butterfly's body (the same way flavor compounds accumulate in a chicken's body), and when a bird eats a monarch butterfly, it gets sick and vomits and forms the same kind of aversion that goats form to blackbrush. That's why a monarch butterfly is brightly colored—to a bird, it looks like nausea. A bird needs to eat a just one monarch butterfly before nutritional wisdom sets in. The alkaloid that acts as a "strategic" compound for the plant acts the very same way in the butterfly.

An even better example is the most toxic substance of all time. This substance didn't rear its terrible head until just over two billion years ago, but once it showed itself, it started killing. Entire species died off, thanks to this poison that was chillingly indifferent to life, triggering a mass extinction. We can only imagine just how badly primitive life on this planet would have been ravaged if not for a fundamental and world-changing evolutionary advance: tolerance. Some creatures evolved a means of handling this toxin. They produced elaborate chemicals all through their bodies that kept this toxic compound in check. Other creatures went even farther. They not only tolerated this substance, they thrived on it. Over millennia, this substance underwent the most stunning PR makeover in the earth's history. The most toxic substance of all time became the most crucial. Cabbage worms, goats, and humans can't live without it. Hospitals store it in enormous tanks. You wouldn't have made it through the last few pages without it.

The substance is oxygen. And it's still toxic. Our cells are littered with oxygen defenses called "antioxidants"—vitamin C is a big one. But they don't always work, and when that happens, oxygen can damage DNA and make a healthy cell cancerous.

Plant toxins are something like oxygen in this respect. The idea that plants hold hidden powers to sustain and heal is as old as history. All those stories you read in newspapers and fitness magazines about cancer-fighting properties of broccoli are, in fact, stories about plant secondary compounds. A postworkout antioxidant power smoothie is a postworkout infusion of secondary compounds.

For quite a long time now, nutritionists have known that diets high in fruits and vegetables and whole grains are healthy. People live longer and don't get as many heart attacks, diabetes, strokes, or cancer. The question has always been, Why? Is it because fruits and vegetables are high in things like fiber and vitamins and low in fat? Or do all those plant chemicals have something to do with it? Over time, the case for plant chemicals started looking pretty good.

As far back as the 1950s, scientists noted that diseases of aging and lifestyle all shared a common trait: oxidative damage (oxygen toxicity, basically). A lot of plant secondary compounds were antioxidants. Could plant compounds help prevent these diseases? The evidence started piling up. In laboratories and test tubes, plant compounds made rats live longer and caused tumor cells to shrivel up. Inflammation went away. It didn't take long before various people were struck with what they thought was an excellent idea: Let's turn these antioxidants into pills.

That's where things started falling apart. Most of the time, antioxidant pills didn't do much. Sometimes, it looked like they did do something, in which case a larger study with more people was undertaken, only then the effect would disappear. And in some cases, the studies went badly wrong. That's what happened with beta-carotene, the most famous member of a family of plant compounds called carotenoids. A study was undertaken to see if beta-carotene and vitamin A pills could reduce the risk of cancer in smokers and people who work with asbestos. They didn't. Not only did the pills fail to prevent cancer, they seemed to cause more cancer, not to mention more death and more cardiovascular disease. The results were so alarming the study had to be brought to a screeching halt. The most recent pill to wash up on the Island of Failed Antioxidants is resveratrol, a secondary compound found in red wine and peanuts that made yeasts and flies live to a ripe old age and caused obese mice to run marathon-like distances, but never lived up to the hype in humans.

Fred Provenza took it all in from the mountains of Utah with a raised eyebrow. What, he wondered, made beta-carotene and resveratrol so special that they deserved to be turned into pills? The foods that contained them—berries, peanuts, grapes, chocolate—featured all sorts of plant secondary compounds. Maybe it was one of the other 44,998 known plant chemicals that prevented cancer or heart disease. For all we know, these chemicals could be acting together in some yet-to-be-understood way. To Provenza, dumping one or two antioxidants in a pill seemed like a kind of blind gamble. It was like ripping apart the Boston Symphony Orchestra into individual musicians, pointing at the Polish gentleman who plays second violin, and saying, "It's him! He's the reason the music sounds so good."

There was also the question of dosage. Even if beta-carotene or resveratrol was good for you, why was a big dose better? That's not how it worked in nature. Goats and sheep, he knew, had a delicate relationship with plants and their many chemicals. A goat, for example, might take a break from chomping ryegrass to nibble on Mormon tea, a shrub that contains a secondary compound similar to ephedrine. But the goat nibbled only a bit. Just because a few Mormon tea leaves were good for a goat, it did not logically follow that the goat should swallow the entire bush.

And just what, by the way, was a goat doing by nibbling Mormon tea or any of the other forty-nine plants on a typical day's menu? A goat doesn't eat a weird plant because her friend's roommate's Pilates instructor says it prevents cancer. Goats fulfill nutritional needs. Was it possible that all these plants contained compounds that goats and sheep, in their nutritional wisdom, benefited from in some way?

It was time for more experiments. Provenza, who is now professor emeritus in the Department of Wildland Resources at Utah State University, spent that latter part of his career exploring the strange and unexpected ways sheep use secondary compounds.

Sheep, Provenza learned, were more like wooly little pharmacists. Take bitterbrush and sagebrush. Bitterbrush is a shrub with a tannin in it that makes it mildly toxic. A sheep can eat some but not too much. Sagebrush is a shrub with a terpene in it that's much more toxic, especially at certain times of year—Provenza once fed a sheep too much sagebrush terpene and it passed out and had a seizure. (It recovered.) A sheep could not and did not eat too much bitterbrush, and it sure kept its distance from sagebrush. But it turned out that in a sheep's stomach, the tannin in bitterbrush binds to the terpene in sagebrush with the strange result that two toxic plants were much less toxic when eaten together. Even stranger, sheep could learn this. They went out and nibbled on bitterbrush, and then moseyed over to the sagebrush and nibbled on it. They ate in courses. The flavors had to go in a certain order for the meal to work. Very quickly, the peculiar grazing practices of French shepherds didn't seem peculiar anymore.

In another experiment, Provenza fed sheep a choice of either plain grape pomace or grape pomace mixed with tannin extracted from quebracho (a kind of tree) known to have antiparasitic qualities.* When the sheep were healthy, they tasted the pomace with quebracho tannin and said no thanks. Then Provenza infected all the sheep with a parasite and offered half of them the tannin-treated pomace. These sheep's tannin consumption shot up. Quebracho tannin made them feel better. Very quickly, quebracho tannin became the thing to eat. Their parasite load, meanwhile, went down.

The implications were huge. Plant toxins were more than simply "poisonous." They could make food more nutritious. They could improve a sheep's state of health. Similar results were rolling in from other fields of science. When tiger moth caterpillars were infected by parasites, they developed a taste for plants containing alkaloids toxic to those parasites. European songbirds preferred food with secondary compounds called flavonoids, and the flavonoids gave their immune system a boost. Honeybees responded to a fungal infection by consuming plant resins. After becoming infected by parasites, chimpanzees chewed a plant called bitter leaf, and a day later they felt better.

The palate had more than just a PhD in nutrition. It had an MD. It

---

* Pomace is what's left over when juice is pressed from grapes. It has very little nutritional value, so sheep don't ordinarily like it; hence, it provides a good test of medicinal effects.

could guide sheep toward food that had medicinal qualities. Provenza proved it. He devised an experiment that showed sick sheep preferred the flavor associated with the medicine they needed. Sheep really could eat their way to good health.

ON A spring afternoon not long ago, Provenza and I were driving up the side of a Utah mountain near the Green Valley Ecology Center. The lawns were neat and trim, and the homes got newer as we made our way farther up the mountain. Eventually, he pulled onto a gravel road and parked next to a bouncing creek that plunged into a pool of frigid water. We got out and started up a dirt path.

During his years at the Green Canyon Ecology Center, Provenza had examined nutritional wisdom in laboratory conditions. Now he wanted me to see how it all worked in nature. We followed a trail along a game fence, a twelve-foot wall of metal mesh that cuts across the slope to keep deer from wandering into Logan to ravage gardens and make dogs go mental, stopping here and there to examine vegetation. Bromegrass. Bluebunch wheatgrass. Arrowleaf balsamroot. At last, Provenza found the shrub he was looking for: sagebrush. This was a young plant, maybe a foot off the ground, with last year's dried yellow blossoms pointed to the sky like smiling faces. Provenza gestured up the mountain. "Do you see much sagebrush up there?" There was none, only this one little plant. Then Provenza pointed to the other side of the game fence, where sagebrush was everywhere.

The difference was deer. Deer eat bitterbrush, which detoxifies the ordinarily toxic sagebrush, just as it does for sheep. Where there were deer, there wasn't much sagebrush; where there were no deer, there was lots.

The bitterbrush was locked in a chemical war with sagebrush. And deer were its henchmen, paid in calories. On this side of the fence, the bitterbrush was winning. Over on the other side, where there were no deer, sagebrush was crushing bitterbrush.

I was experiencing the stirrings of ecological wisdom. A moment ago, the valley was a springtime wonderland of birdsong and blossoming plants nibbled by creatures with fuzzy ears. Now it had gestalted into a war zone, psychedelic in its blend of familiarity and weirdness. All these pretty creatures trying to hurt one another. There were allies and enemies and enemies of enemies that were allies. I was looking at a mountain covered in chemical factories. Some of the chemicals were essential nutrients, some were poisons, and many were somewhere in between. How was a goat or a deer or a caterpillar supposed to survive in such a dangerous place?

The answer was sitting in the middle of my face. Nature's $350,000 instant gas chromatograph. The nose and mouth. Flavor.

I tasted some sagebrush. There was a heady blast of citronella, sage, and pepper, but the leaves were also green, astringent, and bitter. I was sensing millions of years of chemical strategy.

I was not tasting everything, however. Just as a mere handful of chemicals can evoke pure vanilla extract, only a small fraction of the hundreds of secondary compounds in a plant are flavorful. This is by design. Flavor is information. Compounds like phosphorus and vitamin C are stable. They don't waft off food. So the body senses what it can—those unstable floaty aromas—and associates them with the post-ingestive effects on our bodies. "Flavors are like labels," Provenza told me, chemical tags an organism uses to identify and remember the food it has eaten.

Flavor is a chemical language, nature's mother tongue. Bitterbrush "talks" to goats through tannins. Cotton leaves "talk" to wasps through terpenes, sesquiterpenes, and alcohols. Fruit entices fruit eaters with its chemical bouquet. Even bacteria have a rudimentary ability to sense chemicals. But if flavor is a language, it's a peculiar one. The words all mean something different depending on whether you're a goat or a caterpillar, if you have a parasite, and what your body knows. To a human, sagebrush is pungent and bitter. To a sheep or a deer, it pairs beautifully with bitterbrush. To a parasitic wasp, cis-3-hexenol means, "The cater-pillar is over here," to a plant it means, "We are under attack," and to a human it is a crucial note in the label for "strawberry."

There was still something I didn't get. How did a sheep understand it "needed" a particular mineral? A sheep doesn't know it's deficient in phosphorus. A sheep has no idea there's a tannin in bitterbrush that binds to a terpene in sagebrush. How does it "know" what to eat and what it should avoid?

Provenza answered with a single word. During the uncountable hours I'd spent reading books and journal articles with titles like "Common Cellular and Molecular Mechanisms in Obesity and Drug Addiction" and talking to flavor scientists and brain scientists about volatile compounds, taste and aroma receptors, and the parts of the brain that light up as pleasure happens, I had heard this word only once. It was back at McCormick's Technical Innovation Center after I'd reached a state not far off intoxication from Saigon cinnamon. On this point, Fred Provenza and Marianne Gillette were in complete agreement. Flavor all comes down to one thing: feelings.

Animals *desire* particular foods. A sheep infected with a parasite experiences a need state for a flavor that will make it energetic and lively once again. A pregnant goat craves flavors that will bring its fetus protein. The quest for deliciousness is the fuel that powers the behavior, the god that breathes life into the machine. Animals eat what they need because what they need tastes good.

Provenza described the excitement you see in cattle, goats, and sheep when, after a winter of nothing but hay, the snow melts and green shoots start pushing their way out of the ground. "One of the things you notice about animals eating good food," Provenza continued, "is that they're very content." Goats' refusal of young blackbrush shoots, furthermore, is outright. They want nothing to do with it. Provenza pointed at his hand, then his arm and body, and said, "Every organ and every cell has receptors similar to what's in your nose and on your tongue." Creatures communicate within their environment the same way they communicate within their own bodies—through chemical trigger substances that bind to receptors and produce responses. "It's all part of a feedback system," Provenza said, "that tells the body what's good and what isn't."

Goats are not stupid after all. They don't bumble through the world eating what they were born to like. They experience need states. They sniff, taste, and swallow chemicals. They feel curiosity, cravings, satisfaction, and delight along with aversions so strong a mere hint of something can make them turn away in disgust. Flavor is what nutrition feels like to a goat.

If goats had a word for delicious, it would have two meanings. The first would be: I like this. The second would be: This is what my body needs. For goats, they are the same thing.

SIX

Bait and Switch

ARE HUMANS are nutritional idiots?

The mere existence of Doritos, cronuts, deep-fried Twinkie burgers, sizzle sauce, bacon-themed desserts, cheddar smoked sausages, cheesecake, and all the other treats fed to those food-addicted rats in Florida, the fact that there is even such a thing as the Yale Food Addiction Scale, that 69 percent of Americans are obese or overweight, that the American Heart Association now recommends bariatric surgery as a "viable option" for patients who are severely obese, that Debbie continues eating certain foods "even though I am no longer hungry" four or more times a week, that five-year-olds are getting diabetes because they eat too much, and that there is a plus-size coffin maker in Indiana called Goliath Casket would all appear to attest to one irrefutable fact. Our palates aren't just out of tune with our bodily needs. Our palates are out to kill us.

When the movie *Sideways* came out in 2004, merlot became famously untrendy. Nearly overnight, a wine preference became a wine aversion. It had nothing to do with biology. This popular buddy movie didn't somehow cause a toxic alkaloid to appear in bottles of merlot. The change that happened was in people's heads. Scientists have witnessed these changes as they happen. In a now notorious study conducted at the California Institute of Technology, people reported experiencing more pleasure from the same wine when they were told it was more expensive. The pleasure was genuine. As the subjects sipped, an fMRI scanner recorded their brains in action, and the pleasure parts lit up like suburban Christmas lights. (Thus demonstrating a correlation between nutritional idiocy and economic idiocy.)

Nutritional idiocy follows its own demented logic. Humans do not bumble through the world and eat the wrong food by accident. We like the wrong food. In laboratory experiments very much like the ones Fred Provenza did with sheep, humans have shown over and over again that they, too, form "learned preferences" for flavors—when they're paired with calories. We like the flavor of bacon—and cupcakes, french fries, Doritos, potato chips, ice cream, and pizza—because of the post-ingestive association with calories. Scientists have even witnessed these flavor preferences in action in brain scans.

Provenza, as it happens, says animals like calories, too, for the very simple reason they wouldn't survive without them. But here the agreement ends, which brings us to one of the more puzzling differences of opinion in all of science. In one camp, we have the animal scientists and ecologists who, with Provenza, see animal food choices as intimately linked to nutrition. (One biologist described it to me as "foundational.") In the other camp, there are the scientists who study humans, who sound a lot like Provenza's old colleague who laughed at the stupidity of goats. For them, the whole idea of "eating to fulfill needs" or "the wisdom body" is hocus-pocus. Humans eat for one reason and one reason only: calories, calories, calories.

Let's call this the calorie zombie theory of deliciousness. Humans are programmed for "dietary energy." We stagger through our environment on a never-ending quest for calories and just bumble into vitamins and minerals along the way. As a column in the *New York Times* recently put it, "We eat because everywhere we look there's a superabundance of food and we're hardwired through evolution to keep our body weight up." Obesity, by that logic, is destiny. We're doomed.

And yet.

And yet despite nutritional idiocy that has reached pandemic proportions in the developed world and is threatening to become the number one deadliest preventable disease, there are rays—bright, illuminating rays—of hope. Humans have, on occasion, demonstrated fleeting instances of what looks a lot like nutritional wisdom. Really.

Humans experience aversions just like goats. For example, if you ate a rancid crab cake in 1994—as I did—and sprayed crab cake slurry at a high velocity onto the wall-to-wall carpeting covering your bedroom floor at 4 a.m., you too would have developed an aversion to crab cakes that held on until the mid-2000s. You could look at pictures of crab cakes and watch 3-D movies of people eating crab cakes and feel just fine. But the slightest whiff of crab cake would bring waves of nausea.

When humans become deficient in water, they crave liquid refreshment (otherwise known as thirst). When humans become deficient in salt, they crave salt. During the late nineteenth century, explorers in the Canadian arctic surviving on lean wild hares ate more protein than their bodies could tolerate. As they became progressively sicker from protein poisoning—diarrhea, nausea, headaches—they began craving the one substance that could most quickly correct this nutritional imbalance: fat. Pregnant women in the tropics are often overcome by a powerful desire to eat mud, clay, or dirt. The practice is known as geophagy, and all sorts of animals, including Fred Provenza's sheep and Amazonian parrots, do it. These dirt-eating women face considerable scorn, but they do it anyway. Evidence suggests eating dirt can correct a mineral imbalance and help a body rid itself of toxins. (Toxins that are harmless to adults can cause developmental abnormalities in fetuses.) It doesn't happen only in the tropics. Craveworthy "white dirt of Georgia" can be purchased online for $5 a bag.

In 1926, a Chicago pediatrician by the name of Clara Davis undertook one of the most amazing experiments in the annals of nutritional research when she persuaded several teenage mothers and widows to place their infants in her care for six years. Fifteen babies, ranging in age from six to eleven months and who'd never been exposed to "the ordinary foods of adult life," were put on an experimental diet in which they could eat whatever they wanted so long as whatever they wanted appeared on a list of thirty-four foodstuffs that included water, potatoes, cornmeal, barley, beef, lamb, bone jelly, carrots, turnips, haddock, peaches, apples, fish, orange juice, bananas, brains, milk, and cabbage. The foods were all "natural food materials." There was no sugar, no cream, butter, or cheese, and no potato chips, but there was salt for sprinkling. Each item was presented over the course of a single day. The experiment measured "self-selection." Children were presented with the food but in no way encouraged to eat this or that. If they wanted to eat with their fingers, no problem. What they ate and how much was up to them.

The prevailing scientific view at the time was that children were guilty of the gravest nutritional idiocy. Frantic mothers pleaded with doctors about children who wouldn't eat their vegetables. The leading doctors of the day advised that these children be starved until they did. So Dr. Davis set out to discover what babies transitioning from breast milk to food would eat if it was all left up to them.

The answer: everything. At first, anyway. During the initial two weeks, children sampled a little of all thirty-four foods. (This is exactly what goats would do, according to Fred Provenza.) But over time, they each developed favorites, although these would change suddenly and unpredictably. There were generalities—the children came to prefer protein from milk, meat, liver, and kidney, for example, over vegetable protein. And some meals were strikingly unconventional. One child had a pint of orange juice and liver for breakfast. Another had eggs, bananas, and milk for dinner.

Taken as a whole, however, the children chose remarkably balanced diets. They "throve," as Davis put it. Constipation was "unknown." Colds lasted for only three days. When the children were growing and needed protein, their protein intake shot up. When the growing slowed and activity increased, their energy intake increased. During the one "epidemic"—an outbreak of "acute glandular fever of Pfeiffer" (now called mononucleosis) during which every child "came down like ninepins"— there was a curious spike in the consumption of raw beef, carrots, and beets as the children convalesced. Several babies began the study in poor condition. Four were undernourished and three had rickets, a vitamin D deficiency. The very first infant Davis received, in fact, had a severe case of rickets and with each meal was given a small glass of cod liver oil, which contains vitamin D. Children's hatred of cod liver is legendary, but this child consumed it "irregularly and in varying amounts" of his own free will until he was better, then never touched another drop.

These children, Davis found, were master nutritionists. By the end of the study, their overall state of health was so good that another pediatrician, one Dr. Joseph Brennemann, called them "the finest group of specimens from the physical and behavior standpoint that I have ever seen in children that age."

One of the foods those children loved best was fruit. When you stop to consider all the fiber and vitamins in fruit, fruit sure seems like a great example of nutritional wisdom. And which fruit do we like best? Ripe fruit, of course—which contains more sugar than unripe fruit. So really, liking ripe fruit is exactly what you'd expect from a race of calorie zombies.

Not so fast. Fruit's moment of peak edibility isn't simply about sugar. As a fruit ripens, its skin surrounds an eruption of minerals, color, and aroma compounds. The mix of secondary compounds switches from unpleasant and repellant to attractive. Fruit becomes floral and fruity. But then, as ripeness drifts into overripeness, vitamins, acids, and secondary compounds decline, and balance is lost, and the fruit becomes so cloyingly sweet it's nearly offensive.

The difference between good fruit and bad fruit isn't the simple story of sugar. It comes down to secondary compounds. For example, Ataulfo mangoes, those golden-fleshed lobes of joy that Mexicans rightly regard as king among mangoes, have a density of secondary compounds that's roughly double that of regular mangoes (and way more vitamin C). The tiny "wild" blueberries that taste so much more intense than those oversized broiler-chicken blueberries similarly pack more secondary compounds. Generally speaking, the more flavor there is in fruit and vegetables, the greater the density of secondary compounds. It's true of tomatoes. It's true of grapes. It's true of strawberries. It's true of carrots. This should in no way come as a surprise. It is the opposite of dilution.

If the human love of fruit is about more than just sugar, one might reasonably wonder if there is some example of humans consuming plant secondary compounds without sugar.

As it happens, there is a large and very old industry worth billions of dollars that cranks out megaconcentrations of secondary compounds. This industry has been waging a long and largely successful war against dilution. Growers prefer old plant varieties that don't produce much fruit but whose fruit is denser in secondary compounds. They plant these varieties on the hillsides so that they may "struggle," a condition that mobilizes plant defenses and fills the fruit with secondary compounds. And halfway through the season, farmers walk up and down their rows of plants and snip off perfectly good fruit so that the remaining fruit, a pathetically small crop, is even more chemically puissant. The fruit is picked, crushed, and turned into a drink with very little sugar in it that's sold in bottles. Devotees of this beverage get into discussions about how a bottle from this patch of land tastes better than that bottle from that patch of land only a hundred feet away. The differences they discern, however pretentious they sound, are real and come down to plant secondary compounds. The more alluring those compounds, the more this liquid costs.

Perhaps you have heard of this unusual beverage. It is called "wine."

Finally, there is the matter of leaf eating. Humans don't nibble Mormon tea leaves the way goats do, but they have been known to make tea from them. (It used to be served in Nevada brothels.) Humans make tea from sagebrush as well as from a secondary-compound-rich plant abundant in Asia that is so popular as a source of tea people just call it "tea." Tea is a steaming cup of secondary compounds. And humans do eat some leaves. Lettuce, for example. The oddest thing about human leaf consumption is that some of the leaves we like best are richer in secondary compounds than the leaves goats eat. They are chemical bombs. Humans also seek out rare kinds of bark and seeds that are similarly profuse with plant chemicals. I am not referring to obscure, wincingly bitter herbal remedies from the Amazon sold in stores by people who think fluoride is a conspiracy. I'm talking here about a class of food that Americans ate a billion pounds of last year. Archaeologists tell us humans have been adding these strange and potent plants to their cuisine for at least six thousand years. (They have even been found in the teeth of Neanderthals.) We love them. We have fought wars over them. Hank Kaestner spent a career purchasing these plants by the hundreds and thousands of tons. I am talking about herbs and spices.

The human interest in food—outside psychology and nutrition departments, at least—sure looks like it's about a lot more than calories. If we just wanted calories, after all, what's the point of grating nutmeg? Why toss a few sage leaves or a clove into the roasting pot? The last thing a race of calorie zombies should find delicious is a chemically potent plant.

The standard explanation is that herbs and spices are a holdover from times past, when they were used to preserve meat. This makes a bit of sense—spices do have antimicrobial properties—but not much. It doesn't explain why spices were used in vegetarian dishes, or why, hundreds of years ago, European aristocrats from cool countries like England and the Netherlands, where there wasn't nearly the spoilage risk there is in hot climates, paid ridiculous sums for exotic tropical spices to sprinkle on their food. Or why, in an age of refrigeration, we cling to this apparently outdated practice when so many other outdated food practices—gruel, soured milk, mead, salt beef, stale bread, flat beer—have receded into history. If spice use is some peculiar historical hangover, why is it up by 500 percent since 1918?

All of which poses an obvious question: Do herbs and spices have post-ingestive effects? Could they be good for us?

Coriander, it just so happens, "inhibits pro-inflammatory mediator expression by suppressing NF-kappaB activation and MAPK signal transduction pathway in LPS-induced macrophages." Fennel extract exhibits "inhibitory effects against acute and subacute inflammatory diseases." Ginger alleviates nausea and vomiting in pregnant women, kills cancer cells, and can help regulate blood pressure. Dill promotes skin elasticity. Basil kills viruses and prevents inflammation and lowers cholesterol in hyperlipidemic rats; cinnamon decreases blood glucose in people with type 2 diabetes; black pepper exhibits antidepressant properties and stimulates the digestive tract; cloves modulate platelet activity, and elderly Singaporeans who eat curry with turmeric (the stuff that makes curry powder yellow) have better cognitive function than those that don't. It is also thought that turmeric may be antiparasitic and cardioprotective and possess anticancer properties. It can be downright challenging to find a herb or spice that is not a raging antioxidant and does not have some degree of hostility to cancer cells or bacteria.

The effects are not large—and that is the point. If they were large, you would need a prescription from your doctor to visit the spice aisle of the supermarket. People don't take herbs and spices the way they take pills. They take them in low doses over a period of decades. So the bigger question is, What is the effect of all these little doses over a lifetime?

Before we attempt to answer this nearly unanswerable question, let's remember why plants produce secondary compounds in the first place: biological potency. Plants make them so that they can do something to another living organism—kill bacteria, repel goats or insects, attract honeybees, warn a friend, and so on. They do stuff to living bodies. Their healthful effects in humans, however, are not well understood, in part because things in nature like coriander and basil can't be patented so there isn't a lot of money being thrown at them, and in part because long-term studies that measure small effects of low doses are expensive and don't yield the kind of unambiguous, major effects you get with pharmaceuticals, but mainly because preventions are never as exciting as cures. Of the dozens of scientists I spoke to over the course of researching this book, however, not a single one doubted that the peculiar chemicals plants make are an important reason fruits and vegetables are good for you. The whole reason scientists started turning them into pills, after all, was to capture their healthful effects in a bottle.

ONE OF the most compelling stories of nutritional wisdom took place on June 10, 1741, when a sixty-gun British tall ship called the *Centurion* limped into anchorage on the north shore of a South Pacific island called Juan Fernández. Two months earlier, as the gale-battered ship rounded the southern tip of South America, a terrible sickness had set in. The crew was overcome by fatigue, "shiverings," "tremblings," and "putrid fever." In April, forty-three men had died at sea and in May that number nearly doubled. Strangest of all, long forgotten wounds became bloody again. A sailor who'd been injured fifty years earlier during battle in Ireland found that one of his bones, mended for half a century, was once again fractured, and his wounds "broke out afresh and appeared as if they had never been healed." By the time the ship made landfall, more than two hundred men had succumbed.

The *Centurion*'s crew was suffering from scurvy, a malady we now know to be due to a lack of vitamin C. In the eighteenth century, the cause was anyone's guess. Theories ran the gamut from too much salt, to an excess of acid in the blood, boredom, and the body's longing for soil. The palate, however, knew the cure. "In our distressed situation," the

Centurion's chaplain wrote, "languishing as we were for the land and its vegetable productions (an inclination constantly attending every stage of the sea-scurvy), it is scarcely credible with what eagerness and transport we viewed the shore." On Juan Fernández, the crew found watercress and purslane, "excellent" wild sorrel, "a vast profusion of turnips and Sicilian radishes," as well as a moss that smelled like garlic. This curious collection of foraged items was "extremely grateful to our palates."

According to Jonathan Lamb, a historian of voyaging at Vanderbilt University who could well publish a book called *Tales of Nutritional Wisdom on the High Seas*, an acute desire to eat fruits and vegetables was a characteristic symptom of scurvy. As one contemporary naval physician wrote, a sufferer could "gather strength even from the sight of the fruit." Upon eating it, "the spirits are exhilarated, and the juice is swallowed, with emotions of the most voluptuous luxury." The Scottish physician James Lind, who proved scurvy could be cured by citrus fruit by carrying out the first clinical trial of the modern era, sounds almost as though he was channeling goats in his treatise on the disease. "Nature points out the remedy," he instructs. "The ignorant sailor, and the most learned physician, will equally long, with the most craving anxiety, for green vegetables, and the fresh fruits of the earth."

As scurvy progressed, its victims would be overcome by yearnings for food, for home, and for land so forceful that sailors would retreat into a state of wretched homesickness and wake from their dreams in tears. This condition was called "scorbutic nostalgia," and it gets its name from the same root as the technical name for vitamin C, ascorbic acid. A nutritional deficiency we now remember for its hideous physical symptoms—ulcers, black skin, foul breath, and what Lamb describes as "a strange plethora of gum tissue sprouting out of the mouth"—was famed in its day as a disease of *wanting*. Entire crews, far from home and starving on naval provisions, were waylaid by incentive salience. Scurvy, you could say, is the ultimate need state.

A more recent tale of nutritional wisdom begins in the early 1990s with a psychobiologist named Gary Beauchamp. Beauchamp, who for twenty-four years was the director of the Monell Chemical Senses Center and has spent his career studying taste and smell, was contacted by the British makers of a cold and flu remedy called Lemsip, who had a flavor problem. The remedy had been recently reformulated with the anti-inflammatory ibuprofen—the stuff in Advil—and customers were upset. The hot, lemony beverage they previously adored now tasted bitter, they claimed. Beauchamp gathered some colleagues and sat down to a tasting of pure ibuprofen. The Lemsip customers, they discovered, were wrong. Ibuprofen didn't taste bitter. It caused a peculiar burning sensation. It was like the burn inflicted by hot peppers, but it occurred only at the back of the throat.

Beauchamp's colleagues suggested some ways the Lemsip makers might reduce the burning throat problem and Gary Beauchamp didn't think much about the taste of ibuprofen until 1999, when he attended a conference in the medieval town of Erice, which is plunked on a mountain high above Sicily's northwestern coast. Two physicists from Palermo had brought their own olive oil to the conference and hosted a tasting. The oil was poured into little glasses, and guests sniffed, sipped, aerated, and swallowed, just like at a wine tasting. And when Beauchamp swallowed his sample, "Lo and behold," he recalls, "there was exactly the same throat irritation I got from ibuprofen."

In less than a minute, a hypothesis was raging through Gary Beauchamp's mind: olive oil contained an anti-inflammatory. The anti-inflammatory operated on the same metabolic pathways as ibuprofen. And this anti-inflammatory had something to do with the famous health benefits of the Mediterranean Diet.

Back at the Monell Center, Beauchamp poured a glass of olive oil for a colleague who tasted it and said, "Why did you put ibuprofen in it?" In time, it was confirmed that olive oil did indeed contain an anti-inflammatory—a defensive plant secondary compound dubbed oleocanthal—that stimulates a receptor found in the human throat and operates on the very same inflammation pathways as ibuprofen. In a paper published in *Nature*, Beauchamp postulated that constant low level doses of oleocanthal in the Mediterranean Diet may be responsible for some of its health benefits, such as lower levels of cancer, heart disease, and Alzheimer's.

The best part about oleocanthal, however, is how people feel about it. Novice olive oil users don't like it—the stuff makes your throat itch. To connoisseurs, however, throat burning is a mark of quality. Olive oil lovers rate oils on a one-, two-, or three-cough scale (more coughs are good). The European Union considers throat burning an official attribute of fine olive oil. It all sounds an awful lot like nutritional wisdom. Fine olive oil is what scientists call a learned flavor preference and what gastronomes call an "acquired taste." You don't like it the first time, but once the body perceives a health benefit, the oil is consciously experienced as delicious. Throat burn tastes good because it's healthy.*

Speaking of deliciousness, if humans really are calorie zombies, then shouldn't Big Macs, ice cream bars, soft drinks, and the caramel fountain at Golden Corral be the very pinnacle of culinary gratification? And rich people should all be fat because, as the calorie zombies with biggest wallets, they can afford the most calories. (Statistically, they are skinnier.) The restaurants that serve the most expensive and, one presumes, the most pleasurable food are not filled with extraordinarily obese clientele in the throes of epic food binges. Fine restaurants feature

---

* Beauchamp is a fountain of good flavor stories, and here's another: Years ago Beauchamp was involved in the development of a patented method of feeding chickens spices to give chicken meat a pleasing flavor. He approached the CEO of a large chicken company to sell the idea of better-tasting chickens, and the CEO told him why it could never work. Unwanted parts of chickens, he told Beauchamp, are added to chicken feed and then fed back to chickens, and so spice-fed chickens would be fed to regular chickens and eventually there would be no way of differentiating spice-fed chicken from regular chicken. Then the CEO said this, "If you could give me a flavoring that could make chickens taste like chicken, that I would buy."

trim diners, a good deal of whom do not seem to be in it just for the calories. They order small pieces of raw oily fish that, it just so happens, feature brain-healthy omega-3s. They relish just-picked asparagus, say, or sautéed langoustine next to pearly drops of emulsified oyster sprinkled with crumbled seaweed. As they eat these expensive small portions, they do not sit there silently fending off cravings for stuffed-crust pizza and bottomless Dr Pepper. Given the choice between oily raw fish and stuffed-crust pizza, a striking percentage would opt for the fish. The line cooks in fine restaurants—men and women who have devoted themselves to the pursuit of gustatory joy—have unfettered access to food in the top 1 percent of delectability, and yet, strangely, they keep their consumption in check. If it's corpulence you want, you won't find much of it at a restaurant with a three-month wait for reservations. You will find it at Denny's.

This brings us back to Italy. How is it that the country with arguably the most delicious food in the world, where the passion for eating is such that feuds erupt over recipe disputes, is not the foremost hot zone of obesity? (The similarly trim Japanese are also renowned as fine eaters, and the French, too.) The much loved cuisine of Italy features all sorts of foods that throw a wrench into the calorie-zombie hypothesis, like bitter *aperitivi*, bitter radicchio, capers, olives, cured fish roe (*bottarga*), and olive oil with so much throat burn that sipping it makes you talk like the Godfather. Italy is full of food that doesn't taste good the first time you try it but becomes an acquired taste. Italy, furthermore, is home to the diet with the most stellar track record of nutritional research behind it. During the last few years, there has been a small avalanche of high-quality studies attesting to the enduring healthfulness of the Mediterranean Diet—which is not, it should be pointed out, a contrived "diet" in the lose-twenty-pounds-in-two-weeks! sense but a traditional way of eating first described three years before the Chicken of Tomorrow contest and ten years before the first commercial gas chromatograph. The Mediterranean Diet features less obesity, prevents heart disease, stroke, type 2 diabetes, and chronic illness in the elderly and reduces chronic inflammation. Plus, it tastes very, very good. It somehow achieves deliciousness without resorting to dose creep, without constantly upping the hit of sugar, fat, salt, carbs, and umami, and without lacing every meal in flavor chemicals.

HUMANS ARE a lot like goats. We have similar chemical sensors—noses and tongues—and the same internal feedback mechanisms. The problem with humans is all that equipment doesn't seem to be working.

Back at the Green Canyon Ecology Center, Provenza showed me why. He produced a clear plastic baggie containing an off-white powder with a street value of $5. It was Sucram, a palatant so potent this 255-gram (9-ounce) bag could flavor more than a ton of livestock feed. Provenza spread the bag open so his research colleague, Juan Villalba, and I could each stick our noses in. I took a whiff and promptly lost peripheral vision. Imagine a tanker truck filled with butterscotch sauce, gooey caramel, and candy floss. Now imagine that tanker truck getting hit by an eighty-megaton bomb. I wet the end of my finger, dabbed it in palatant, and touched it to the end of my tongue. If a kindergarten class completed a hostile takeover of a flavor company, this is the substance they would create.

It was time to introduce these sheep to the Dorito Effect. Six sheep would get a choice: ground hay and/or ground hay mixed with Sucram. As he explained the methodology—dose, duration, and so forth—unknown to Provenza, a tiny spout of Sucram was flowing out of a pin-sized hole in the bag. The aroma surrounded us like mosquitoes in spring bog. We began waving our arms. I coughed up a pearl of mucus that tasted like salted caramel.

The sheep loved it. Four out of the six buried their faces in their troughs and all you could see were their furry little ears moving back and forth as they gobbled. At one point, Villalba said, "This is a flavor that looks like an attractant," which is as close as a scientist gets to saying, "Wow." After ten minutes, Provenza and Villalba removed the troughs and weighed their contents. The results were unequivocal. Not only did twice as many sheep prefer the Sucram-treated feed, they ate more of it—15 percent more—than the sheep that preferred regular hay. The two holdouts, furthermore, weren't hardened to the temptations of synthetic flavors. Provenza repeated the experiment, once with apple flavoring and again with umami flavoring. Each sheep, in the end, proved to have a weakness.

It's what's known in fraud circles as bait and switch. The palatant fools the sheep into eating something that is different from what they're actually eating. In the case of Sucram, the sweet, milky flavors remind farm animals of mother's milk. That's why it works so well with piglets. It pushes the same pleasure buttons as the milk that once flowed from their mothers' teats. But obviously it is not mother's milk. Sucram lets farmers peel off the flavor label for "mother's milk" and attach it to something else—hay, corn, soybeans, whatever—all while jacking up the sweetness. Animals think it tastes great.

If this sounds like a recipe for nutritional confusion, the confusion goes deep. Not only does Sucram make unpalatable feed taste better, it improves "feed efficiency." Livestock gain more weight when there's Sucram in their food, a bonus for farmers going for maximum poundage. A study at the University of Liverpool shed some light on the underlying biology. Sweet receptors, it so happens, aren't found only in the mouth; they're in the digestive tract, too. The artificial sweetener in Sucram stimulates these receptors in piglets' guts, and this causes a digestive hormone to be released, which in turn causes the digestive tract to increase the absorption of nutrients.

Is something similar happening with people? Perhaps the best-kept secret of the diet industry is that artificial sweeteners don't seem to be doing us much good. Many studies show that people who consume them suffer from an alarming risk for obesity, metabolic disease, hypertension, stroke, heart disease, and type 2 diabetes. This could, of course, be due to "reverse causation"—the whole reason people use artificial sweeteners is that they are obese or prediabetic to begin with. (But even if that's true, it's pretty clear artificial sweeteners aren't doing much to help.) There is, however, reason to believe the problems may go deeper. In rats, artificial sweeteners have been shown to induce what the behavioral neuroscientist Susan Swithers refers to as "metabolic derangements," in which sweetness loses its "meaning"—the body no longer regards it as a signal of impending calories—causing the rats to eat too much and gain weight. (The very opposite of nutritional wisdom.) The Yale neuroscientist Dana Small hypothesized that if eating artificial sweeteners "erodes the relationship between sweet taste and calories," then that eroded relationship should be measurable in the human brain. In a study similar to Debbie's milk shake brain scan, Small fed subjects sugar and found that a part of the brain called the amygdala—the part where feelings are felt—did indeed light up differently in people who used artificial sweeteners. If that isn't enough cause for concern, a more recent study found that high concentrations of artificial sweeteners can encourage the formation of human fat cells. The lesson is compounded: You can fool the tongue but you can't fool the body.

You can also fool the nose. Our superbly sensitive noses are fooled regularly and often, to great and profitable effect. The most illustrative example of nose fooling comes not from food but from a product whose very purpose is to encourage repeat and escalating dosage: tobacco. Cigarettes are flavored just the way Doritos, potato chips, soft drinks, salad dressings, chicken nuggets, and supermarket cheesecake are flavored. Men and women with advanced degrees purchase flavorings in bulk from flavor companies and put them in cigarettes. It is lucrative employment. A recent job posting by Philip Morris International states, "The successful candidate will play an integral role in the creation and development of cigarette flavor systems" and this person's responsibilities will include "ensuring cigarette flavors and/or flavor application systems are transferable from the laboratory to the flavor production facility and the cigarette production facilities." A flavorist's artistry is assessed by a "trained smoking panel."

As to the question of which flavor chemicals are added to which cigarette, good luck finding that out. They are listed under the same vague umbrella terms "natural" and "artificial" flavorings as happens with food. The R. J. Reynolds Tobacco Company, however, publishes a master list of all the ingredients it puts in its cigarettes. Of the 145 items on the list, 131—90 percent—are flavors. The notes they impart include "nutty," "creamy," "woody," "smooth," and "butterscotch." A 1972 reference document titled "Tobacco Flavoring for Smoking Products" states, "The taste of licorice to the smoker is that of a mellow sweet-woody note which, at proper use levels, greatly enhances the quality of the final product." An industry newsletter from the same year is more blunt about the intended effect: Flavorings "make the product sell better."

These chemicals have lurked in mainstream cigarettes for decades. In recent years, however, tobacco flavorings have had a sort of coming-out party. Smokes now come in nonsubtle, exciting flavors like French grape, chocolate mint, and cherry. And just like young pigs love Sucram, young humans love flavored tobacco. They love it so much that the federal government banned flavored cigarettes in 2009 and now high school smokers have moved to where the flavor is: flavored little cigars, a product that in 1972 the industry recognized "can be consumed in volume because the taste and aroma is acceptable to the young smoker."*

It is classic bait and switch. Cigarette smoke comes cloaked in flavors that seem familiar, wholesome, and natural, and then quietly slips in a

---

*Newly popular e-cigarettes are available in thousands of flavors, including blackberry, cherry, cinnamon, cotton candy, mango, and honey. But the greatest example of flavor-enhanced abuse is the recreational use of a "morphine lollipop" called Actiq. Intended for cancer patients but popular with off-label users, it consists of a lozenge perched on a stick and dosed with fentanyl, an opiate 50–100 times stronger than morphine, and is laced with "artificial berry flavor."

dose of something with a powerful post-ingestive feedback: nicotine. Food is following the identical model. And we keep getting better at it. The process of identifying flavor mixtures in nature and then cranking them out in factories used to take months, even years—in the case of vanilla, a century. Now we can do it in a matter of weeks, if not days. Flavor companies dispatch teams to rain forests and citrus groves to hunt down eye-popping new flavors, and the chemicals are brought back to headquarters for analysis and reproduction. In 2012, the world's largest flavor company, Givaudan, sent an expedition to Mexico that led to the creation of "authentic mango flavours." Givaudan has even developed a Virtual Aroma Synthesizer, a trumpet-shaped device that mixes flavors in real time until the optimal version of "cherry" or "strawberry" is achieved. It is used with "children panels"—flavor focus groups of kids—who engineer a perfect cherry with a few clicks of a mouse. As a Givaudan executive put it to me, "We get a very good idea what children consider the best cherry."

We know exactly what this does to animals. Fred Provenza found sheep that ate hay one day would eat even more hay the next day if a flavoring was added. It created what Provenza calls a false sense of variety. The same effect has been documented numerous times with rats—scientists refer to it as the "variety effect." When scientists in Italy sprayed ryegrass or clover flavors on straw pellets, the goats preferred the flavored pellets (especially the ryegrass). Scientists at Japan's National Grassland Research Institute achieved the same result by spraying flavors on hay, describing these compounds as "intake stimulants." This is called a "generalized" flavor preference, and it, too, has been observed in rats. The implications may seem benign when it comes to goats or sheep eating hay, but consider it in the context of a child. Would a seven-year-old girl be interested in drinking a bottle of water mixed with sugar? The answer is no. (I tried it. The response was, "Daddy, that's just gross. It's way too sweet.") But add some flavorings to sugar water and the child thinks it tastes like juice and will finish a whole bottle.

Flavor factories churn out chemical desire. We spray, squirt, and inject hundreds of millions of pounds of those chemicals on food every year, and then we find ourselves surprised and alarmed that people keep eating. We have become so talented at soaking our food in fakeness that the leading cause of preventable death—smoking—bears a troubling resemblance to the second leading cause of preventable death—obesity.

WHEN THE sheep had finished up the Sucram hay, Provenza and I got back in the car and descended into Logan. We were in search of false variety, and we found some at a drugstore where we stopped to buy batteries for my digital recorder. Two boys were standing at the checkout, scrounging their last nickels to purchase two cans of AriZona fruit drink. At the tender age of perhaps eleven, they already had what a farmer would call "finish" on them, a layer of buttery fat that smoothed their features and made their skin look like it would jiggle like Jell-O in a bowl. They had achieved the state of livestock perfection—young and plump. Their basketball high-tops looked blown out and haggard from withstanding all the heft pounding down on them, and to behold their knees was to imagine a future of joint replacement.

We pulled in to a supermarket next and found intake stimulants, trigger substances, and palatants—which is to say "natural flavor," "artificial flavor," MSG, and the rest of the receptor-stimulating brigade. It was on the first shelf we came to, in the buttercream frosting. It was in the cookies and it was in the strawberry-swirl cheesecake. Provenza picked up a container of blue cookie icing. "This stuff," he said, "even looks toxic."

There was palatant in the hummus, tapioca pudding, honey butter, butter, beef marinade, and fruit and grain bars. "I wonder what happens when you eat blueberries," Provenza said, "and then you eat this?" He was pointing at a box of Western Family Blueberry Waffles, which were next to strawberry waffles and apple-cinnamon waffles, all of them synthetically flavored. I asked Provenza what happens to nutritional wisdom in sheep and goats when the same flavor is paired to two different nutrients. "They get confused," he said.

We were exploring the perimeter, the so-called healthy part of the supermarket, but we still couldn't escape palatant. There was "natural flavor" in the almond milk and soy milk (along with plenty of sugar). "Natural" flavor was in every single yogurt we picked up. It was in Alpo Chop House Originals dog food.

The most breathtaking stretch of false variety in any supermarket is the drinks aisle, where sugar water goes around pretending it's the most nutritious plant in the forest. There were regulars, like 7Up (lemon-lime), Canada Dry (ginger), A&W (sassafras), and Pepsi (kola nut), and the niche players, like Country Time (lemon again). But they were now fighting for shelf space with the New Age drinks: Honest Tea, SoBe, Peace Tea, neuro SONIC (a drink for "mental performance" that contains resveratrol). The AriZona brand drinks staked a prime swath of real estate right at eye level. I asked Provenza to choose one. "No thanks." I insisted. Provenza picked out a bright purple can: "Grapeade."

Talk about bait and switch. A drink whose name starts with "grape" lists filtered water as its first ingredient and high-fructose corn syrup as its second ingredient, and has more pear juice in it than grape juice. The reason it tastes like grape is thanks to ingredient number six: "natural flavors," which very likely contain the grape-evoking compound methyl anthranilate, although it's impossible to know because AriZona, when asked, told me, "That information is proprietary to our recipe." The bait and switch doesn't end at flavor. On the side of the can there was a big logo that said "ANTIOX" in bold font with "vitamin C" and "antioxidant" curling around it in smaller type. To be clear: Vitamin C, which is listed as the last ingredient—and is therefore the least ingredient—is, as the can states, an antioxidant. But it's one antioxidant. In an actual grape there are too many to count.

Out in the parking lot, Provenza handled the can of Grapeade with the nervous unease of a kid who'd stolen a joint from his older brother. "Go on," I said. He clicked it open and gingerly sipped. "If I take one swallow," he said, "I can relate to grape and sweet." He sipped again. "But if I drink half that can, the experience will be negative. The aftereffect on my system will not be good." He walked over to a garbage can and dumped it in, "antioxidants" and all.

That's the problem with Logan and all the cities and town humans now live in. So many of the flavor solutions come from factories. Above the game fence, only blueberries taste like blueberries. Above the game fence, the line between things and their flavors is more like a wall. Below the game fence, anything—waffles, yogurts, drinks—can taste like blueberries. Thanks to dilution, they often taste more like blueberries than actual blueberries. Below the game fence, food whispers irresistible chemical lies.

Imagine, for a moment, that foods are destinations on the highway of nutrition. The flavors are the highways signs—they are how a body knows where it's going—and the place they take you is the actual nutrition. Out there in nature, there's a turnoff with a big sign above it that says grape. If you take that road you'll get to a place with lots of vitamins C and K, some thiamin and riboflavin and potassium, fiber, sugar, and all manner of plant secondary compounds (phenolic acids, flavonoids, tannins, proanthocyanidins, and on it goes). At the supermarket, the same sign—grape—takes your body to Grapeade, a place with no fiber and a bit of vitamin C, but that packs a serious hit of sugar. In nature, a desire for strawberries will take you to a food with vitamins, minerals, fiber, plant compounds, and a bit of sugar. At the supermarket, that desire takes you to a drink called Strawberry Colada that is mostly water and sugar, and to strawberry-flavored waffles and a yogurt snack for kids whose package features enticing pictures of strawberries and a flavor that makes kids feel glad, energetic, free, and confident but that contains no actual strawberries, just synthetic flavoring and sugar.

In nature, flavors all take you to different nutritional destinations. At the supermarket, they all lead to the same place: calories. At the supermarket, the highway of nutrition becomes like some creepy episode of the *Twilight Zone*, where every town you come to looks different—one is Mexican, one is Italian, one is French—and yet beneath each superficial façade they're all exactly the same.

Below the game fence, we keep changing what "delicious" means. For decades now, we have been gently but steadily nudging up food's caloric dose and dressing it up in chemicals that make it seem real and nutritious. It's gotten to the point that it's hard to find food that doesn't feature some degree of bait and switch. Food is becoming more like cigarettes, and the result is predictable. Casual food users have become heavy food users. And heavy food users have become addicts.

So let's review all the ways the Dorito Effect appears to be turning us into nutritional idiots:

▸ Dilution. As real food becomes bland and loses its capacity to please us, we are less inclined to eat it and very often enhance it in ways that further blunt its nutrition.

▸ Nutritional decapitation. When we take flavors from nature, we capture the experience of food but leave the nutrition—the fiber, the vitamins, the minerals, the antioxidants, the plant secondary compounds—behind. In nature, flavor compounds always appear in a nutritional context.

▸ False variety. We naturally crave variety in food—it's one of nature's ways of making sure we get a diverse diet. Fake flavors make foods that are nutritionally very similar seem more different than they actually are.

▸ Cognitive deception. Fake flavors fool the conscious mind. A mother enticed by a Dannon Strawberry Blitz Smoothie as an after-school snack for her eight-year-old child will taste it and reasonably believe the product contains strawberries, even though it contains none.

▸ Emotional deception. Flavor technology manipulates the part of the mind that experiences feelings. Fake flavors take a previously established liking for a real food and apply it, like a sticker, to something else—usually large doses of calories—creating a heightened and nutritionally undeserved level of pleasure.

▸ Flavor-nutrient confusion. By hijacking flavor-nutrient relationships, fake flavors, by their very nature, set a false expectation. A major aspect of obesity is an outsized desire for food, one that very often cannot be extinguished by food itself. By imposing flavors on foods without the corresponding nutrients, are we creating foods that are incapable of satiating the people who eat them? So many of the foods we overconsume—refined carbs, high-fructose corn syrup, sugar, added fat—would not be palatable without synthetic flavor. We gorge on them because they taste like something they are not.

THERE IS one more problem: micronutrients.

So far, we've looked mainly at how flavor is manipulated. But that's only one side of the flavor-nutrient coin. Think back, for a moment, to Fred Provenza's experiment where he paired maple or coconut flavoring with phosphorus. It's easy to see the flavoring as the trick. After all, the sheep grew to love either maple or coconut, even though neither one was responsible for alleviating the phosphorus deficiency. But the experiment was, as Provenza puts it, a "total contrivance." The phosphorus was fooling the sheep just as much as the flavors were. Without the dose of that needed mineral, they would never have formed the attachment to the flavor. There's a lesson here that's easy to miss: Nutrition can drive behavior.

That's how it works with animals. One of the first signs of a vitamin or mineral deficiency in livestock is a reduction in intake. If a goat or cow isn't getting what it needs, it stops eating. The food that formerly tasted good now tastes bad, and the animal begins craving different food, because on some level it knows its regular food is so deficient it has become almost like a toxin. This happens all the time in nature. On the Scottish island of Foula, sheep engage in one of nature's greatest affronts to vegetarianism by eating baby arctic terns. With surgical precision, they clip off wings, legs, and heads, because they need the minerals. Red deer on the Scottish island of Rhum graze on puffins. And caribou on the shores of Hudson Bay have been observed ravaging the nests of snow geese for their eggs and they spit out the fuzzy little chicks because it's the shells they're after, for the calcium. It was a similar nutritional need that moved Provenza's goats to dine on wood rat urine back in 1976 and that moved the scurvy-ravaged sailors on the *Centurion* to gorge on wild sorrel, turnips, and a moss that smelled like garlic back in 1741.

This doesn't happen with humans today because we don't let it. We pop multivitamins, even though years of research data suggest this does nothing to augment health. We also "fortify" foods. We add vitamins to flavored sugar water and call it "vitamin water." We add calcium to flavored sugar water with soy protein and call it "soy milk." You can even buy chocolate milk with fish oil. By law, refined grain products like white bread and sugary cereals must contain thiamin, riboflavin, niacin, iron, and (more recently) folic acid. In the case of flour, the objective was always honorable: to make sure people don't suffer from deficiencies. Since folic acid enrichment was introduced to the North American diet in 1998, neural tube defects have declined by 25 to 50 percent. But scientists are also alarmed by an uptick in incidents of colorectal cancer that started at the same time.

There are, in other words, consequences—consequences we have barely considered. In nature, if a goat isn't getting its phosphorus, it will set off in search of other plants, and when it finds a plant rich in phosphorus, it will form a flavor preference to that plant. The liking will become generalized—it will nibble it periodically and crave it in times of need. In nature, there are sound nutritional reasons to eat lots of different things. By sprinkling micronutrients in our food, we may be doing the very same thing we do when we sprinkle on flavorings: creating a disincentive to seek out genuine variety. Imagine a growing child with a surging thiamin (vitamin B_1) requirement eating a bowl of Froot Loops with Fruity Shaped Marshmallows. This breakfast cereal is just as much of a contrivance as Provenza's experiments, featuring a whole bunch of "enriched" nutrients married to a synthetic flavoring, along with a whack load of sugar and marshmallows. Not only could the added thiamin be making that child form an even stronger flavor preference to a very sugary cereal, the child has no incentive to form an attachment to a food that is naturally rich in thiamin (cauliflower comes to mind). If it wasn't bad enough that fruits and vegetables have lost so much flavor, we may be giving ourselves even more reason not to like them. And if that all seems like too much to swallow, consider that Italy, France and Japan—three countries that prize their vegetables—don't enrich their flour or rice.

There is a clearer way of seeing this. When Christiaan Eijkman noticed chickens that ate only white rice got beriberi and postulated the existence of vitamins, chickens were, from that point, doomed to confinement. In half a century, they'd be crammed into coops and converting feed into flesh at an unheard-of clip. The high-energy diet gets much of the credit, but the high-energy diet at one time would have caused a deficiency. What made it possible was the discovery of micronutrients, which were added, one by one, to chicken feed. Chickens didn't need to eat leaves, turnip tops, or bugs anymore. Whatever the benefits of nutritional fortification and enrichment may be, this much is undeniable: It enables a high-calorie diet. It's the tried and true formula of industrial farming: calories plus micronutrients equals maximum poundage. Add some synthetic flavoring and the stuff even tastes like food.

Humans look just like livestock now. We achieve a state of buttery plumpness before we've even reached sexual maturity. We experience powerful cravings for food that is slowly making us sick. We are like parasitic wasps dropped out of airplanes: programmed to eat the wrong food. We aren't born calorie zombies, but that's what we have become.

PROVENZA SUGGESTED we stop for lunch at a Mexican spot in Logan's old train depot because it's one of the few spots in town that's not a chain restaurant. (This was in fact incorrect. The restaurant is part of a chain called Cafe Sabor.) He was looking for "positive post-ingestive feedback," which is to say a meal that would leave him feeling good. This is, he believes, a rare quality in a restaurant. When Provenza eats fast food, which happens only when he's traveling, he feels overstuffed and "foggy." He believes a high percentage of Americans go through much of their lives this way. We sat on the patio, where a waitress took our order. She returned, it seemed, in less than a minute with tamales that were perfectly formed and tacos that tasted re-thermalized. The salad needed lots of dressing. Textbook dilution. Flavor solution.

Back in the car again, I broke the post-ingestive silence. "How do you feel?"

"Not very good," he said.

SEVEN

Fried Chicken Saved My Life!
(But Can It Save Yours?)

A MORNING IN September, sunny and clear. A thirty-nine-year-old man with advanced bed head walks downstairs for breakfast when a strange thing happens. His coffee tastes too sweet. His tongue is coated in a kind of cloying sugariness. He can't seem to taste the coffee—only sweet. Did he put in too much sugar? The man doesn't think so, but he brews another cup just the same because coffee is one of the minor high points of his day. When the second cup is ready, he carefully adds a splash of 2 percent milk and a smaller splash of 10 percent cream and more carefully stirs in a single spoonful of sugar.

There it is again. The coffee is too sweet. If his mouth were a TV, the man would slap the top to try and knock it back to working order. He takes another sip and a thought occurs to him. Drink it without sugar.

The man does not take his coffee without sugar. Without sugar, coffee tastes like what it is: roasted bean water. If the man ever accidentally sipped his wife's coffee—she takes hers with no sugar—he'd scrunch up his face and wonder how she could tolerate this thin, bitter soup. The man doesn't worry about calories because he is trim. He feels justified, furthermore, in his sugar usage because many years earlier he had taken something called the PROP test. He placed a piece of paper dipped in a special chemical on his tongue and found that it was distressingly bitter. This confirmed he was "sensitive" to bitter tastes. He needed sugar in his coffee to cancel out all that bitterness. He spoke to one flavor expert who told him that putting sugar in his coffee was a mark of a sensitive palate. He was a proud sugar user. His taste buds were set to maximum.

Something, clearly, has changed. Intellectually, the man is still pro-sugar. His tongue is still anti-bitter. But his feelings now are anti-sugar.

THAT MAN was me.

And the coffee episode was the latest in a series of small but notable changes my palate had undergone over the past many months. The fruit cravings had started coming on about a year earlier. A sudden yearning for fruit was not completely new. Every summer since I was fourteen I'd gorge on plums as soon as they came into season, and as far back as I can remember I have adored the Moroccan clementines—the more sour the better—that mark the beginning of the holiday season. But now I was being seized by spontaneous fruit cravings almost daily. A craving for kiwis lasted several weeks and ended only when kiwis went out of season and I couldn't get my hands on them. I switched from green grapes to red grapes and was eating close to a pound a week. During cold winter nights in February, I would camp out on the couch to watch NHL hockey and down at least one, but more often two and, once, three grapefruits.

The fruit cravings were followed by something that, years earlier, would have been unthinkable: a singular desire for vegetables. As I was making quesadillas for my kids one night, my daughter asked, "What are you having, Daddy?" My answer surprised me more than it did her: an entire head of broccoli, stir-fried with garlic, chilies, and sea salt. At one time in my life I would have happily dumped the global supply of Brussels sprouts into a burbling volcano. Now I was eating Brussels sprouts as often as twice a week. I heard my wife saying theoretically impossible things to me like, "Didn't we just have collard greens two nights ago?" My desire for stomach-stretching loads of sweet or savory was fading away. In the language of McCormick, I was swept up by altogether different need states. I wanted food that made me feel cleansed, energetic, recharged, and clear.

And then, my moment of Nidetchian transformation: I lost weight.

Not much weight—ten pounds. But I wasn't trying to lose weight. I didn't need to lose weight. My BMI went from 24.1 (normal) down to 22.8 (still normal). The secret to this unintended miracle diet? Flavor. Real flavor. Flavor that grows out of the ground or walks off a farm—not the stuff engineered in some nameless lab. And one of those flavors was fried chicken.

IT WAS (and remains) the best fried chicken I have ever tasted. This was not a rational measurement of chicken quality based on crispiness, juiciness, and "fried chicken flavor" the way it happens in a sensory evaluation. This was about feelings. This was get-up-out-of-your-chair-and-start-dancing fried chicken. It brought on a state of giddiness that verged on the evangelical. You know chicken is good when you want to run down the street telling strangers about it.

I was not alone in my fervor. A friend of my wife sent me an e-mail with the subject "Chicken!" four days after eating said fried chicken. (She had never before communicated to me by e-mail, and she has not since.) "I'd like to tell you directly that the chicken on Saturday night is the stuff dreams are made of." Her husband accused me of "ruining" chicken for him, because no future chicken could possibly measure up. The most unexpected thing about that chicken, however, wasn't how it tasted. It was how little I ate.

The road to chicken enlightenment began almost exactly two years earlier with an e-mail from one Douglas Hayes of Calistoga, California. I had just published my first book, *Steak: One Man's Search for the World's Tastiest Piece of Beef*, which was just like it sounds. Over tens of thousands of miles and hundreds of pounds of beef, I had come to the unexpected and uplifting conclusion that the most flavorful steak is also the healthiest and comes from cows that eat only grass. I'd spent my life thinking the opposite, that delicious things were by their very nature bad for you. And now here was something delicious that was not. Strangers were sending e-mails to congratulate me on getting things right. Others—almost always people with a financial stake in feeding grain to cows—sent e-mails to tell me I was an idiot. Mr. Hayes of Calistoga was in a class all his own. He wrote to praise a meat I had openly disparaged in my book due to its paucity of flavor: chicken. Naturally, I gave Mr. Hayes a call.

Hayes is the kind of character that, like certain types of grape, thrives in Northern California. He did two tours in Vietnam and was headed for a PhD in quantum mechanics but changed his mind and became a builder, then a contractor, and later an architect, and spent nearly three decades doing mainly residential and restaurant design inspired by the organic architecture of Frank Lloyd Wright and Bruce Goff. By 2008, as Hayes was tiring of architecture, he became involved in the Buckeye Recovery Project. The Buckeye, he told me, is a rare breed of chicken that peaked in popularity at the turn of the last century at a population of between two and three million birds, but was down to about eight hundred by the time Hayes caught wind of it.

"What do these chickens taste like?"

"You can't imagine," Hayes told me. (I imagined chicken. I imagined biting into one of those pillows they hand out on airplanes.)

I had given up on chicken because it was too bland. Cooking it was hard. Modern chicken needed honey, soy sauce, garlic, and smoked paprika. A chef I know told me he never cooked chicken without brining it first. Chicken was like a terrible date: high maintenance *and* boring.

What I wanted in food was flavor. This had nothing to do with concern over dilution or an effort to gain nutritional wisdom—I hadn't heard of either concept yet. My motives were purely hedonistic. I wanted to eat food that tasted better. Over a strip loin in a Paris bistro, the famous chef Alain Ducasse upended everything I'd thought about cuisine when he told me that cooking is the easy part. The most difficult thing for a chef, he said slowly, as though intoning Scripture, is procuring the most delicious ingredients possible. Ducasse was describing what's known as "ingredient-driven" cuisine. I became preoccupied—my apologies here to Monsieur Ducasse—with Italian cuisine, which eschews culinary complexity and prizes the food you're cooking, not the stuff you put on the food you're cooking. Following the lead of Marcella Hazan (a lady who bragged about not owning a spice rack), I grew to mistrust spices, thinking that all they did was enable bland food. (My fundamentalist stance on spices has since softened.) I became skeptical of the new generation of dude chefs, the tattooed twenty-somethings who brine and smoke everything and then roll it in panko and deep-fry it, blitz it with harissa, and then cover it with bacon and maple syrup and melted cheese and Nutella or some other low-hanging flavor hit beloved by children and potheads. They were giving their ingredients the Dorito treatment. But their ingredients were bland, so what choice did they have?

I wanted food that was like that steak I had come to love, food that tasted strongly of itself. I combed farmers' markets for carrots that tasted the most like carrots, potatoes that were the most potatoey, and peaches that tasted the peachiest. There just wasn't any room in this new flavor order for chicken. Chicken didn't taste like itself. Chicken didn't taste like anything. It tasted like what you put on it. And even then, the flavor of what you put on it would play out too fast, like gum. Once you got past the skin, you had to contend with the actual chicken meat, which had as much flavor as a hotel bathrobe. What Hayes proposed was radical. Chicken actually could have flavor.

Hayes put me on to an even more central figure in the heritage chicken movement, a man named Don Schrider, who told me that chicken flavor, though very real, was nearly extinct. Schrider described a lost world of chicken, a wonderland of colored feathers, sizes, shapes, and ages. Certain breeds, he said, were legendary in their day for the way they tasted—the Dorking, the Crèvecoeur, La Flèche. Were we talking about the same bird here? Long ago, the male offspring of egg-laying chickens were not gassed or ground up. They were raised to be "fryers." Up until the late 1930s, Schrider told me, the Pullman Coach Company served slow-fried chicken in its railroad dining cars from a laying breed called a brown leghorn. (The *Looney Tunes* character Foghorn Leghorn is the most famous member of this breed.) I was struck by a nostalgia for a time I'd never experienced. I fantasized about taking the train from Chicago to San Francisco in 1922, not so that could visit these great cities in their Roaring Twenties prime, but so I could hunker down in the dining car and eat fried chicken.

In terms of flavor, Schrider said there was just no comparing a modern broiler. "It's not just the way they taste," he said. "Heritage chickens make your tummy happy." Schrider, for the record, was unfamiliar with the concept of "post-ingestive feedback" or the work of Fred Provenza. He was just talking about feelings.

Schrider talked up another breed, the barred rock—the variety that induced chicken-and-dumpling tears of joy in that Wichita wife. I found a hatchery not far from Toronto popular with Mennonite farmers that had barred rock chicks for sale and arranged for a hundred or so hatchlings to be delivered to a friend's farm. After a few weeks on starter feed, they goose-stepped into the greenery like they owned the place and began pecking in their terrifically precise way at blades of grass, slugs, frogs, seeds and, once, a snake. At fourteen weeks old, they were more than double the age of supermarket broilers and around half the size. They were entering the now vanished phase of poultry delectability called the "fryer" stage, a prescription as much as a description. One afternoon, we killed the biggest barred rock in my friend's flock. Beneath the great puff of black and white feathers was something small and bony. The skin was yellow and the fat inside the carcass was darker yellow. It didn't look like chicken. It looked like a dead bird.

There were only two and a half pounds of dead bird, bones included, so my wife and I waited till the kids went to bed to eat it. I got out my favorite cookbook of the salt-and-pepper era, 1902's *Ideal Cook Book.* This, of course, is when the knock comes at the door. It was my wife's friend. She was moving to New York the next day with her husband and they dropped in to say a final goodbye.

Come in, come in. How about a beer? But won't you stay for dinner? Won't you stay for some dead bird?

The plan was as follows: Fill up on corn on the cob. Fill up on salad. Because there wasn't enough dead bird for four adults. Our guests entered the kitchen and politely eyed the puny main course, which looked like I'd strangled it in its nest. I cut the fourteen-week-old barred rock into pieces, seasoned them with salt and pepper, dragged them through flour, and fried them in a cast iron pan filled with a shallow layer of hot oil. When the chicken was brown and crispy on all sides, I poured a sip of water into the pan, slammed a lid on top, let it steam for twenty minutes, removed the lid, and cranked the flame back up. The chicken was now golden brown and crispy. And even smaller than before.

And my, what a chicken. The stages of pleasure went as follows: incredulity, astonishment, elation, glowing thankfulness. There were declarations featuring the words "awesome," "best," and "unbelievable." We were drunk on chicken. Most amazing of all, somehow there were leftovers. I put two small pieces in the fridge for tomorrow. There was enough meat to make at least one more tummy very happy.

TWO MONTHS later I visited one of this continent's more storied fried chicken spots, a Memphis joint that has been dunking breasts and drumsticks into batter and then hot fat for decades. However old the restaurant was, the broilers it was serving up were ultramodern—plump, meaty, and bland. The exterior perfectly executed the Dorito model: crispy, fatty, and loaded with MSG. The actual meat beneath it was dry as toilet paper. And yet I couldn't seem to stop eating it. I didn't dislike the chicken, but my level of liking was in no way commensurate with the pace at which I was eating it. After every swallow, I wanted more. It was as though the chicken created an itch that only more chicken could scratch. I ate till the bucket was empty. I was stuffed.

One afternoon, I was at a birthday party with my four-year-old twins (one girl, one boy) when something similar happened. As the children gathered around the hired children's entertainer to sing "Five Little Monkeys Jumping on a Bed," I made a move to the snack table, where I encountered a blue plastic bowl filled with those unmistakable powdery orange triangles: Doritos. I told myself I would have precisely one. This, of course, proved to be impossible. I crunched the single Dorito, and the sound of children singing faded out as a singular craving took hold. The want for more Doritos got so strong it was painful. I pried meager chunks of compressed Dorito out of my molars with the tip of my tongue and launched them down my throat, but it was like trying to fill the Grand Canyon with a fistful of pebbles. I cleansed my hand of seasoning, as though purity of body would bring about purity of mind. It was no use. Within a minute, I reached back in the bowl for more. The analytical part of my brain sat back and watched the primitive zones of desire and reward play their cat-and-mouse game.

Just like with the Memphis fried chicken, I didn't enjoy the Doritos, not the way I enjoy good chocolate or a ripe peach. The only payoff was the first bite—an entertaining crunch and a brief spike of flavor that all faded to mush. Desperate to regain the thrill, I would reach in and grab another chip. It wasn't that putting a Dorito in my mouth felt good. It was more like not eating them felt bad. As the children moved on to "Bumblebee, Bumblebee Fly Away," I was loading them in four at a time.

It was a moment out of the Yale Food Addiction Scale. I find myself continuing to consume certain foods even though I am no longer hungry. It happened again at a McDonald's in northern Vermont, where, on a family road trip, we pulled off the highway and I ordered a Big Mac, Coke, and medium fries, and downed all 1,120 calories in maybe three minutes. I wanted more. My wife said wait, that in twenty minutes my brain would register that my stomach was full, but I could not heed this wise advice. She took the kids to the bathroom and I seized my chance, ordering a cheeseburger and small fries. As I stood there pushing the soft, salty food in my mouth, I watched two obese couples waiting to place their orders in the line that had now formed, impatient, frustrated, their incentive salience roaring. My mood crashed. I became disgusted with the food, with myself, and with everyone around me. I threw the half-eaten cheeseburger and what remained of the fries in the garbage and stormed into the bathroom to wash the grease off my hands. In the car, I experienced Yale Food Addiction Scale symptom #5: I spent a lot of time feeling sluggish or fatigued from overeating, a condition my wife began referring to as McRegret.

There was never any McRegret with heirloom fried chicken, just love and happiness—and unexpectedly small portions. These dead birds were impossible to find at supermarkets and butcher shops, so I made friends with farmers who could provide my heritage chicken fix. I would order a year's worth of poultry in February or March and, come July, place them in my new chest freezer. In time, I learned how to kill a chicken without severing its spinal cord, which makes it easier to pluck. When I would tell my children we were having chicken for dinner, my six-year-old daughter would ask, "Daddy, did you kill it yourself?" and when I answered in the affirmative she'd clap her hands and say "yesssss." I served heritage fried chicken to one of the moms from my kids' school. She cocked her head back, as though struck by a thought, and said, "This is what Kentucky Fried Chicken is trying to taste like." I gave some to the chef who brines his chickens, and he said, "It tastes the way chicken used to taste." My father concurred. He bit into a leg and said, "I haven't had chicken like that since just after the war." Chicken from another era transported people to that era.

This didn't make sense. With the exception of my father, who was born in 1934 and has lived and eaten during the prebroiler era, these were identifiably false chicken memories. The mom and the chef had never actually eaten the chicken they were apparently remembering. They'd both grown up on broilers. Is this, I wondered, the opposite of food addiction? Is this what happens when food delivers a reward that is greater than the anticipation?

Something was up with this chicken.

AND GOOD LUCK finding out what. If ever there was such a thing as a golden age of chicken flavor science, it is long over. Back in 1954, scientists Marion Sweetman and Ingeborg MacKellar posed the important question: Do male chickens taste different from females? The answer was yes, they taste stronger. Four years later, science confirmed that the difference extended to chicken soup. (This is consistent with the traditional French view that old roosters make the best coq au vin.) In 1962, twenty-eight-month-old chickens were found to be more flavorful than nineteen- and nine-week-old chickens. And in 1965, it was found that "muscle from young and old chicken had similar chemical constituents, but the older ones had larger concentrations."

As chickens lost flavor, however, poultry science stopped studying flavor. It was about broilers, now. It was about quantity, efficiency, and price. If the question of flavor came up, it wasn't about which chicken tastes the best. It was about whether or not a potentially very cheap feed like fish meal or canola meal might make a broiler taste weird. Flavor didn't matter. That's what flavor solutions were for.

The number of poultry scientists today specializing in chicken flavor can be counted on a single hand. One of them is Linda Farmer, who works at Belfast's Agri-Food and Biosciences Institute. Farmer told me that despite the shortage of research, we can be sure that chickens acquire flavor in two ways. The first is through diet. Simply put, what chickens eat ends up in their bodies. This is why modern broiler farmers add a yellow pigment to chicken feed—so the chickens on the supermarket shelf will look the same as chickens that have been running around outside eating leaves. As to exactly what delectable substances are making their way into the flesh of, say, a fourteen-week-old pasture-raised barred rock? No one cares. Poultry scientists are interested in broilers, and broilers don't eat pasture.

There is also the issue of age. In 1997, Farmer did a study that concurred with *Mrs. Norton's Cookbook*—older chickens have more flavor. Exactly why they have more flavor—which compounds accumulate in a chicken versus which ones get filtered out—remains a mystery no one is keen to solve because there's no research money in it. Modern broilers, after all, are flavorless babies.

When it comes to nutrition, the picture is clearer. In 2009, a seminal figure in British nutrition named Michael Crawford, of London's Institute of Brain Chemistry and Human Nutrition, compared chickens of today with chickens of times past. If the inquiry sounds similar to the work Donald Davis did on modern fruits and vegetables, it was. And just as with fruits and vegetables, the results were troubling. Humans aren't the only ones who are way fatter than they used to be. Back in 1870, a 100-gram (3½-ounce) piece of chicken contained just under 4 grams of fat. By 1970, that number had risen to 8.6 grams, and by 2004 chicken was packing more than 20 grams. As Crawford dryly noted, "While chicken was at one time a lean, low-fat food, it is no longer." Less dryly, he asked, "Does eating obesity cause obesity in the consumer?"

Not only is there a lot of fat in modern chickens, it's a bad kind of fat. Chicken today contains much more omega-6 fat than it used to and much less omega-3 fat. To Crawford and his coauthors, this is unsettling. Both of these fats are considered "essential"—without them, you die. However, they must be consumed in balance, and an overabundance of omega-6 fats, which is typical of modern diets, is associated with arthritis and cancer as well as inflammation. The most important omega-3 fat is called DHA, which is believed to be good for both the brain and heart. We are encouraged to eat oily coldwater fish such as salmon and mackerel because their flesh contains plenty of DHA. Cattle, pigs, and chickens, we are told, don't have any DHA, only a much simpler omega-3 called ALA. Chickens, however, have the ability to turn significant quantities of ALA into DHA. They have a biological talent that lets them make their flesh more like mackerel and salmon. At least, they used to. But no more, because we don't feed them much omega-3s, and they don't live long enough to do much converting. "The decline of dietary DHA," Crawford and his coauthors write, "and the disturbance of the omega-6:omega-3 ratio are thought to be contributing to the rise in mental ill health." (The good news, I suppose, is that if bland chicken is giving us Alzheimer's, we'll all eventually forget how bland it tasted.)

It all comes down to what chickens eat. Or don't eat. Study after study has confirmed this. If they eat grass or chia, their meat contains more omega-3s. If you plunk laying hens on pasture, their eggs contain more omega-3s (but production of said eggs dips, which is why hens don't often see the likes of pasture). I brought up the issue of omega-3s to Linda Farmer, and here she had something very interesting to say. Omega-3s are flavorful—in beef, at least. These immensely unsaturated fats are highly prone to forming aromatic compounds in a frying pan or on a hot grill. It is a case of nutrition you can directly taste. Is the same true with omega-3s in chicken? "I think it is likely no one has looked at the flavor compounds in chicken reared in a completely different way," Farmer said. Everyone is into broilers.

Pastured eggs also have more vitamins E and A. The yolks are brighter, which comes down to a family of plant secondary compounds called carotenoids (which also accumulate in a chicken's liver, shanks, and skin). In a poultry research project that someone ought to turn into a made-for-TV movie, chickens allowed to range on the Tibetan plateau, where "there was a dense population of grasshoppers," ate the Tibetan grasshoppers and, as a result, had more vitamin E and iron in their meat (not to mention more antioxidants and better shelf life) than chickens cooped up in cages fed the high-energy diet. A chicken is what it eats, pure and simple. The more green vegetable matter it eats, the more its flesh becomes incrementally more like a green vegetable.

I was struck by a thought. If the research on real chicken flavor was lacking, did fake chicken flavor have something to tell us? Do the powders we sprinkle on chicken offer any clues about what's missing in chicken? I was thinking, specifically, of kokumi, the little-known Japanese taste sensation that improves the continuity, mouthfulness, and thickness of salty, sweet, and umami even though it has no taste of its own. According to Ajinomoto, one of the most important triggers of the kokumi sensation is a protein called glutathione. Why, I wondered, do humans like glutathione? Why would evolution have endowed us with the ability to taste glutathione and be pleased by it?

Not because a human needs it. Glutathione isn't like salt or vitamin C—something you must eat to survive. You can make all the glutathione you need (so long as you aren't protein deficient, in which case glutathione is just one of your problems).

I contacted Xingen Lei, an expert on oxidative stress at Cornell University who studies glutathione. "You could say it's the most important antioxidant," he said. "It's the first line of defense against oxidative stress." Glutathione is of such fundamental metabolic importance that Dr. Lei doesn't know of a single cell that does not contain it. Glutathione is in animals, it's in plants, it is even in anaerobic bacteria that don't breathe oxygen. I asked Dr. Lei what would happen if all the glutathione were to suddenly disappear from your body. You would experience "redox collapse," he said, and die. (Note to self: Avoid redox collapse.)

Perhaps glutathione functioned as a kind of indicator of quality? Maybe the "mouthfulness" and "continuity" in the flavor of a fourteen-week-old barred rock indicated the bird was vigorous and free of disease? "It's reasonable," Lei said. "If you are healthy, you should have normal levels. If you have a compromised condition, you will have a diminished amount."

A poultry scientist I spoke with measured just that—diminished amounts of glutathione—in a group of chickens with metabolic disease. A study from Argentina found that cattle fed a lot of grain also have diminished glutathione. Pasture-fed cattle, in contrast, not only had more glutathione in their cells, they also had more antioxidants from plants (vitamins C and E and beta-carotene). And those grasshopper-eating chickens on the Tibetan plateau? They were similarly rich in glutathione. Are kokumi flavorings another bait and switch? Is modern chicken bland in part because chickens are metabolically stressed? And by sprinkling on kokumi seasonings, are we furnishing their meat with the illusion of health? No one has looked.

Scientists have looked at glutathione in plants, however. I spoke to Tom Leustek of Rutgers University's department of plant biology and pathology, who told me glutathione is often considered an indicator of freshness in produce. For example, if Japanese plums are harvested late and then stored too long, the result is "chilling injury"—brown spots, mealy flesh, and a generally unpleasant eating experience. With chilling injury, glutathione levels crater, as do levels of vitamin C and other plant secondary compounds. It's as though the fruit has burned through its storehouse of health-maintaining chemicals and the tongue is all too aware of it. In this case, at least, glutathione very much does look like an indicator of fruit and vegetable quality. Leustek called it a "reasonable hypothesis."

So did Linda Farmer. All three scientists used the same wording: "reasonable hypothesis." It means, "This is an interesting question and someone should see what the answer is." And if it's true, kokumi flavorings are just the latest way we have given bad food the illusion of nutrition without actually delivering it.

NONE OF THIS shed any light on the question of portion size. When I ate Doritos or a Big Mac, I kept on eating and eating, and later experienced McRegret. So why when I ate a fourteen-week-old barred rock or a grapefruit did I find it tremendously delicious and yet tremendously satisfying? If these foods tasted better, shouldn't I have just kept on gorging?

Fred Provenza believes the difference comes down to what he calls "deep satiety." "Fundamentally," he told me, "eating too much is an inability to satiate." When food meets needs at "multiple levels," it provides a feeling of "completeness" and offers a satisfaction that's altogether different from being stuffed.

Provenza began thinking about this when he did his free-choice experiment with cattle—the study where he let some calves pick and choose what they ate while others got the premixed feed put together by a professional ruminant nutritionist. The study was notable, of course, because calves did a better job of meeting their nutritional needs than the nutritionist. But it was notable to Provenza for a second reason. The free-choice calves ate less. Despite a bottomless trough of feed in front of them, they didn't keep on eating and eating. That fourteen-week-old barred rock, the thinking went, satisfied me because it was more than a mere vehicle for fat and protein. It contained other stuff my body needed—vitamins, minerals, and omega-3s—and somehow my body knew it.

It also contained secondary compounds. And it will come as no surprise that Provenza believes plant compounds have a lot to do with deep satiety, even when those compounds are in dead birds.

In 2005, he took four lambs and drenched their stomachs with a terpene in sagebrush and then presented them with a meal that featured no added terpene. The sheep ate less. They gobbled for several minutes, then drew their snouts out of the trough and hunkered down while four other sheep—sheep that did not receive a terpene drenching—kept on merrily chomping. The terpene-treated sheep, it is important to note, were not sick. They were bright-eyed and perky, and their ears weren't drooping. They just weren't hungry anymore. Lunch was over. It was as if some kind of internal register said, "Enough." The hunger light went off.

Provenza is not the only scientist to have made such an observation. In a similar study, rats that consumed blueberry extract before their daily meal weren't as hungry as rats that consumed water—they ate less and weighed significantly less by the end of the study. (Scientists refer to plant compounds that reduce appetite as "antifeedants.") The phenomenon has even been observed by Pancosma, the company that makes Sucram. Pancosma sells an oregano extract that, when added to pig feed, kills harmful gut bacteria and promotes intestinal health. The downside is that if you add too much, the pigs end up eating less. (Provenza observed the same phenomenon in sheep. Early in his career, he used oregano as a flavoring in experiments but had to switch to a simple synthetic flavoring because the oregano was making the sheep eat less and blurring results.)

And it works that way with people. Raisins boost the level of a gut hormone associated with satiety. Several studies have found that eating chili peppers seems to reduce appetite and weight gain. Flavor's effects do not end at the mouth and nose. They have only just begun. There are taste receptors all through the digestive tract exactly like the ones in your mouth. Smell receptors are sensing down there. The digestive tract is not some blind extractor of nutrients. It has sensors the mouth and nose lack—fat sensors, protein sensors, bacteria sensors, hormone sensors, even plant-compound sensors. The gut is its own little chemical-sensing gourmand, tasting each bite and adjusting its processes accordingly. You don't taste what your gut tastes, but it does affect your feelings.

To Provenza, this is nutritional wisdom in action. The digestive tract is the battleground where plants exert their evolutionary will over the animals that eat them. And the way they do it is through toxic chemicals. Mother Nature is not trying to poison you. Satiety is reached before you consume a toxic dose. Just like with drugs that a doctor prescribes, when it comes to secondary compounds there is a healthy dose and there is an overdose. Any animal that eats plants has to be able to make sure it doesn't get too much of something. In some cases, the threshold is low—goats nibble only a little Mormon tea, just as humans sprinkle a little oregano into a tomato sauce. But Provenza thinks it also happens with fruit, the food plants so often "want" animals to eat, because fruits are loaded with secondary compounds.

Strangest of all, plant defenses—the very substances a plant deploys to get us to stop eating—often taste extraordinarily good. Consider onions. In their whole state, they possess very little in the way of flavor. It's only when you chop or bite an onion that its characteristic flavor is instantly manufactured, as cell walls are compromised and enzymes kick into action. This is chemical strategy at its best. Injured onion flesh produces compounds toxic enough they can kill dogs and cats. Humans love them. We love the spicy oil pressed from olives, which produce oleocanthal so that animals won't eat them. We like myristicin, a flavor compound in nutmeg and parsley that, in large enough doses, causes headaches, palpitations, and nausea. We are all psychopaths when it comes to eating vegetables. Not only do we register their chemical screams and calls for help, we are delighted by them. If humans could get together with caterpillars, bees, and goats, we'd all raise our glasses and agree on the following: Some pesticides are truly delectable.

It doesn't seem to make much sense. Why do we relish the very chemicals plants deploy to get us to stop eating?

According to one novel theory, that's how we are programmed. Before you dismiss it, it's worth considering a startling yet obvious point: Some of the most intensely pleasurable substances known to humankind are toxic pesticides. Nicotine, cocaine, heroin, and THC—even caffeine—all evolved to "interfere with neuronal signaling in herbivores." And the fact that we get pleasure from being intoxicated—a word that contains "toxic"—is no coincidence. According to this theory, articulated in a paper titled "Explaining Human Recreational Use of Pesticides" and published in the academic journal *Frontiers in Psychiatry*, the human brain evolved a system to regulate the consumption of toxins for the very simple reason that it had to. Any animal that goes around eating plants, after all, needs to make sure it doesn't accidentally kill itself over lunch. And over hundreds of millions of years of evolution, organisms have found curious uses for many of these plant poisons. As much as we justifiably malign nicotine, it is so effective in controlling parasites that some veterinarians use it as a dewormer. Toxins can be very good for us. We just have to be sure we don't get too much.

That's the part we humans are unexpectedly good at—not too much. Nicotine is so toxic a thirty-milligram dose can kill a person inside of five minutes, but smokers expertly suck in fraction-of-a-milligram puffs, titrating their dose to maintain a safe and satisfying blood level. Cocaine is so toxic it causes anxiety and paranoia, so cocaine users snort a line or two, keeping the dose below "acute toxicity," just as a chef does with oregano or habañero peppers. When it comes to the use of plant toxins, humans—just like sheep, chimpanzees, and honeybees—are natural-born pharmacists.

It all got me thinking about my own fruit bingeing. It had undeniably druglike aspects. Sudden cravings, uninhibited consumption, reward by the ladleful. And yet it's hard to see eating plums, clementines, and peaches as abuse. A three-grapefruit bender adds up to about 250 calories, or about 10 percent of my recommended daily intake. A two-kiwi binge is 85 calories, and coffee with cream and no sugar is around 30. These food experiences barely register on the Yale Food Addiction Scale. There is no regret, no "dysphoria" when I'm not using. I don't feel sluggish or fatigued and never melt into a puddle of self-loathing. I don't lie to friends and family about these foods. I invite them over for heirloom fried chicken. And yet, as with cocaine and nicotine, I seemed to titrate my dose. I would eat enough grapefruit—or tomatoes or fried chicken—and I was done. They hit the spot so perfectly that the spot vanished.

Whatever is going on, this much is obvious: There are highly pleasurable foods that do not follow the Dorito model of intake; you don't just keep eating and eating. However fantastic wild blueberries and sweet juicy peaches are, you can't pig out on them the way you can on Big Macs, chicken nuggets, and soft drinks. These foods trigger a deeper, more complete satiety, and the reason, as odd as it may sound, has something do with toxicity. The food nature makes has a lower toxicity threshold than the food humans make. You can't eat quite as much before the hunger light goes off.

(Most of the time. A veteran magazine publisher I know went over the threshold in the late 1960s when, as a junior ad salesman, he failed to close a big deal one afternoon in the quiet but prosperous small city of London, Ontario. On the drive back to Toronto, he plunked down a bushel of fresh apples on the front seat of his convertible and gorged all the way, tossing apple cores out like a medieval duke, and ended up eating so many apples he got sick, and to this day can't nibble on an apple without breaking out in hives.)

Human nutritionists don't think much about toxicity. They tend to see eating through the lens of nutrients—fat, carbs, vitamins, protein, and so forth. But if you walk from the nutrition department across campus to the ecology department, the thinking is completely different. Ecologists think about toxins all day long. All over nature, animals—whether it's deer, monkeys, or caterpillars—limit their meal size not because they're stuffed and couldn't possibly eat another bite, but because they've hit a secondary compound wall. It's not up to them when they stop eating. It's up to what they're eating.

Recently, human nutritionists have begun to take a few tentative steps toward the other side of campus. They've started looking at what all those taste receptors are doing in the gut and have discovered the following: Bitter compounds—like the ones in grapes, blueberries, and broccoli—release hormones that trigger satiety. Bitterness can turn off the hunger light. It has been proposed that the Italian predilection for a bitter *aperitivo* before dinner is helping Italians stay trim. (To those unaccustomed to this fine tradition, I suggest starting with an Aperol with Prosecco, seltzer, and an orange slice. *Buon appetito.*)

Everything, in the end, is toxic. Even water and oxygen can kill you. It is all a question of dose. And that's the other problem with Doritos, Memphis fried chicken, soft drinks, and other Doritoesque foods: They're too nontoxic. They are so nontoxic that there's nothing limiting intake, so we overconsume them and, over time, all that fat, sugar, and carbs ends up becoming toxic. The calories build up in our bodies and interfere with our circulation, choke off parts of our hearts, wear out our joints, and blow out our pancreases. Obesity and so many of the terrible conditions it causes come down to nothing more than calorie toxicity.

It all goes back to Gottfried Fraenkel, poison-loving cabbage worms, and the word "secondary." Food's role as food was always secondary. The reason animals produced meat wasn't so that some human could eat it. Meat is muscle, which enables movement, and fat, which stores energy. Vegetables are the plant equivalent—structure and nutrient storage. A plant grew fruit or seeds not so that a human could bake pie but so that the plant could reproduce. As humans mastered growing livestock and plants, however, we changed their ecological purpose. Instead of eating living things designed by nature, we started doing the designing. And we got so good at it that food became very different. It's a difference we still don't fully understand but one we can most certainly taste.

IN SEPTEMBER 2002, four Danish scientists began examining grocery receipts. They were not hoarders or fraud artists selling credit card information to hackers in Kiribati. They were just curious about who was drinking wine and who was drinking beer. It may sound like the sort of research escapade a thundering politician would decry as a useless waste of taxpayer dollars, but it was the kind of experiment other scientists describe as "elegant." For years, science had been grappling with the unexplained health benefits of wine—wine drinkers seemed to be more resistant to coronary heart disease and certain kinds of cancer, but no one knew why. Predictably, there was a large-scale effort to rip wine apart in search of whatever plant secondary compound was working its peculiar magic on the human body and turn it into a pill. (Resveratrol was one.) The Danish group came at it from a different angle. They didn't need a gas chromatograph. They needed receipts. They wanted to know what else all those healthy wine drinkers were buying when they visited the supermarket.

They did not suffer from lack of data. All together, they examined 3.5 million transactions from ninety-eight different supermarkets. And they found that wine drinkers didn't shop the same way as beer drinkers. The people who bought wine were more likely to place olives, low-fat cheese, fruits and vegetables, low-fat meat, spices, and tea in their carts. Beer drinkers, on the other hand, were more likely to reach for the chips, ketchup, margarine, sugar, ready-cooked meals, and soft drinks.

Perhaps the good health of wine drinkers isn't caused by wine so much as the fact that wine drinkers like wine in the first place. To these people, deliciousness isn't limited to concentrated bursts of calories. They seem less like calorie zombies and more like those nutritionally wise toddlers back in 1926. The greatest predictor of health, these results suggest, doesn't come down to this or that nutrient. It comes down to what a person finds delicious. You are what you like.

I had come to a similar point of view, though through a very different route: fried chicken. Actually, not just fried chicken. There was also grass-fed steak, fruit bingeing, unsweetened coffee, and my new-found and unexpected—and until now, secret—love of collard greens and Brussels sprouts.

This was epiphany material. I had embarked on a mission to eat the best-tasting food I could find—to enjoy the kind of ingredients Alain Ducasse endeavors to serve to the guests at his restaurants—and the result is my palate underwent some kind of modification. I stopped putting sugar in my coffee. I lost weight. I was no longer punished by post-meal hangovers. Most important of all, the way food tasted changed. It became more powerfully satisfying, as water is most quenching after a long run on a hot day.

This became even clearer at McCormick's Technical Innovation Center. Before my day with Marianne Gillette investigating the world of feelings, flavor trends, and the future of crackers, I sat down with the company's chief science officer, Hamed Faridi. Faridi was talking to me about the relationship between spices and health, an area of research McCormick funds. I asked him why it is that herbs and spices, which people eat because of the way they taste, would have the side benefit of being healthy. Faridi responded with an offhand remark, but it stayed with me. "It could be," he said, "that bland things are inherently un-healthy."

The more I thought about it, the more it made sense. Broilers were bland and they were omega-6 fat bombs, not to mention nutritionally enfeebled compared to heirloom chicken. Doritos were originally bland, which is why people used to dip them in actual food like bean dip and salsa and why they are now covered in chemicals that create the illusion of food. The milk shake injected into Debbie's mouth—just like the food fed to those food-addicted rats—was bland, which is why it needed all that synthetic flavoring. And all those high-yielding tomatoes, carrots, and greens in the produce section are also bland, which is why people either don't eat them or smother them in fatty dressings and sauces that blunt their nutrition. Blandness and our attempts to disguise it is the hidden scourge of our time.

The food problem is a flavor problem. For half a century, we've been making the stuff people should eat—fruits, vegetables, whole grains, unprocessed meats—incrementally less delicious. Meanwhile, we've been making the food people shouldn't eat—chips, fast food, soft drinks, crackers—taste ever more exciting. The result is exactly what you'd expect.

It all boils down to what I will call the Rules of Flavor:

1. Humans are flavor-seeking animals. The pleasure provided by food, which we experience as flavor, is so powerful that only the most strong willed among us can resist it.

2. In nature, there is an intimate connection between flavor and nutrition.

3. Synthetic flavor technology not only breaks that connection, it also confounds it.

SO HOW do we fix this?

Well, we could try to fix rule 3. We could try to make synthetically flavored things more healthy.

We have been trying at least since the introduction of diet soft drinks back in the 1960s. Of all the briefs McCormick's food product development department receives, 30 percent are for "healthy" foods. The food industry is scrambling to quench the need state for "health" and health trends blow in and out faster than flavor trends. There was low fat. There was oat bran. There was probiotic and low carb. Now there's "tropical citrus" vitamin water, which contains 100 percent of your daily required dose of vitamins B_6 and B_{12}, niacin, and pantothenic acid (vitamin B_5), and vitamin C. Now there's "wild berry" neuro SONIC, which "offers a tang on the tongue that is soon followed by increased alertness, focus, and mental zip," and summer citrus berry neuro BLISS, "specially formulated by our team of scientists to help you relax, unwind, and recenter." The big word in the industry right now is unquestionably one of its most annoying, "wellness," a trend that has seen a rise of ingredients like green tea and chai and exploits the widespread myth that the forest wants to heal you. McDonald's recently introduced the Egg White Delight McMuffin, a lower-calorie breakfast sandwich that's part of a new effort toward engineering better fast food (and which features artificial flavor in the egg whites and liquid margarine and "natural" flavor in the Canadian-style bacon). For dessert, enjoy a Snackwell's Devil's Food Cookie Cake, a single serving of which (one cookie) has only 50 calories and almost no salt. We can, and we will, engineer our way to good health.

Well, not really. Because the funny thing about whole wheat Ritz Whole Wheat crackers is that they don't have much whole wheat in them—the lead ingredient is refined flour. Tropical citrus vitamin water should probably be called tropical citrus sugar water because after "reverse osmosis" water, the lead ingredients are crystalline fructose and cane sugar. And all those low-calorie concoctions, however hightech, don't taste good. Ask McDonald's what America thought of the McLean Deluxe. Maybe that's a good thing, because when they do taste good, people often end up eating high-calorie quantities, a phenomenon known as the "Snackwell Effect." Can these foods even be called "healthy"? Perhaps we should think about it this way: If you cut a batch of pharmaceutical-grade cocaine with chai, you could say with some degree of honesty that it is "healthier," "less addictive," and "now with chai!" But would you say it's "good for you"?

Food companies entice you with labels that promise health and vitality and then quietly slip you a massive hit of calories as a means of extracting money from you. But there is a deeper problem: skill. If a food company wanted to create a cracker that's as healthy as, say, a strawberry, would it even know how? A strawberry, to pick just one example of a naturally flavorful thing, is healthy, as Fred Provenza would put it, on multiple levels. A strawberry contains vitamin C, vitamin E, vitamin B_6, biotin, folate, niacin, pantothenic acid, riboflavin, thiamin, calcium, potassium, magnesium, phosphorus, copper, boron, iron, iodine, manganese, molybdenum, zinc, omega-3s, histidine, isoleucine, leucine, phenylalanine, threonine, tryptophan, valine, and fiber—along with anywhere from three to five thousand known plant secondary compounds, including ellagic acid, lutein, zeaxanthin, and beta-carotene, and three hundred or so aromatic compounds, about eighty of which we can pin down.

So let's pretend for a moment—and we are deep in the realm of fantasy here—that this food company could procure all these substances, that they could buy the magnesium, copper, phosphorus, zeaxanthin, and all those other plant compounds, and they could crack open *Allured's Flavor and Fragrance Materials Buyer's Guide* and order all eighty known aroma compounds. Would it?

How much would it cost to add in all those vitamins, minerals, and omega-3s? How much for just half of the known aroma chemicals? Imagine the industrial sprawl of tubes, tanks, and evaporators it would take to produce three thousand plant compounds. That would be a seriously engineered cracker.

Food companies have no idea how to make a cracker that complex. And even if they did, it wouldn't taste any good for the simple reason that humans stink at making things delicious. We are in the flavor Neolithic. We have been at this game for a bit over half a century. We may have knocked off a perfect vanilla, but we can't make food pleasurable without resorting to cheap thrills like fat, sugar, carbs, and MSG. Consider: A sour-cream-and-onion potato chip contains 5.7 calories per gram. A Nacho Cheese Dorito has 5. Honey Mustard Wheat Thins come in at 4.7; Mallomars at 4.4; Popchips—"a chip that has all the flavor, and half the fat"—at 4.2; and a Danonino strawberry-flavored yogurt (with no actual strawberry) that is "power packed with essential nutrients to help little ones grow up healthy and strong" at 1.2.

An actual strawberry? 0.32 calorie per gram.

An actual strawberry is a masterpiece of flavor engineering, the food equivalent of a car that gets five hundred miles per gallon. Humans don't know how to create deliciousness on such a measly calorie budget. Nature has mastered the art of hedonic density—food that maximizes pleasure and minimizes calories, but we aren't anywhere close. We think the problem with processed food is that it's loaded with too many chemicals, but the truth is that it doesn't have nearly enough.

THIS LEAVES rule 2: Leave the flavoring up to nature.

We're already doing that. The fivefold increase in spice consumption over the last hundred years is a mass response to dilution. As humans have been leaching their produce and livestock of flavor, they have been simultaneously harvesting armfuls of leaves, bark, and seeds—armfuls of secondary compounds—to get the flavor back in. It is a spark of nutritional wisdom so bright you practically need a welder's mask. And we're not screwing up herbs and spices the way we screwed up tomatoes. McCormick constantly tests new batches with sophisticated equipment. The company once received some Mediterranean herbs grown in California's San Joaquin Valley that tasted "weak" compared to Mediterranean herbs from the Mediterranean. They went straight into the garbage.

In France, there are rules that stipulate if you want to make a truly great cheese such as Mont d'Or, Morbier, or Compté, you may use milk only from particular cows that graze pasture or eat hay. If you want to make Chevrotin, milk production must be capped at eight hundred kilograms per goat per lactation, so as not to become diluted. (A modern goat dairy can achieve yields more than double.) Locals say you can taste the pasture in the cheese. Locals in Normandy say you can taste the pasture in their butter, which is the color of egg yolks, and locals in Spain say you can taste the difference between Iberian pigs fattened on grass and acorns versus the same pigs fattened on grain. Locals are all talking about the same thing: plant secondary compounds. We are unknowingly mesmerized, all of us, by plants and their chemical strategies.

Fine restaurants hire foragers to comb the countryside and gather secondary compounds. Stinging nettle. Chickweed. Henbit. Some of these restaurants have their own gardens and hire I-gave-up-a-lucrative-career-in-law gardeners who pluck heirloom tomatoes off the vine and snip heirloom varieties of lettuce and pull heirloom radishes, fennel, potatoes, and onions from the earth. If you look very hard, you will even find the occasional restaurant that serves barred rock chickens with yellow skin and yellow fat. The defining culinary buzzwords of the last ten years—"local," "heirloom," "artisanal," "pastured," "foraged"—may evoke thoughts of old-timey farming goodness, but they are all unconscious strategies against dilution.

And it hasn't worked.

Actually, that's not quite right. It has worked, but only a little bit. For those without access to barred rock chickens and pastured pork—which is to say, most people—the flavor trend known as spice rubs has made bland industrial meat not only bearable but pretty tasty. Garlic, rosemary, sage, thyme, paprika, and brown sugar will not only give a thirty-five-day-old broiler flavor, it will also add a dose of healthy plant compounds. There's only so much herbs and spices can do for teddy-bear stuffing, however. They can replace lost flavor with a different flavor. But they can't replace omega-3s, vitamins, and minerals, and all your meat ends up tasting seasoned the way potato chips taste seasoned.

For a month or two every year, I can pick heirloom tomatoes in my tiny backyard. I can watch my children pop slices in their mouths and say, "Daddy! That is the best tomato ever!" During the summer, I can buy them at farmers' markets for $6 a pound. They bruise easily and sometimes the skin is cracked and if you don't eat them within a couple of days they become a $6 a pound rotten mess. But when an heirloom tomato is on, baby, it is on. Once or twice a year, my wife and I can afford to eat at a restaurant that has its own greenhouse and laying hens and serves foraged curiosities from fields and forests. If the heirloom movement is a rejoinder to dilution, it has proved its point. Nature can and does produce ingredients of sublime flavor. Alain Ducasse knows what he's talking about.

This is excellent news for the hedge fund manager ordering the $300 tasting menu. This is good news for "foodies" who've discovered the pleasures of an heirloom radish salad on a Sunday afternoon in August. It's good for people like me who live in cities and can buy a loaf of bread made with the deeply nutty, nutrient-rich, and low-yielding ancient grains enjoyed by the protoagriculturalists that existed ten thousand years ago. But it's not good news for most people most of the time. Because what the heirloom has also proved is that extraordinarily flavorful ingredients are expensive.

The reason, alas, is yield. Even if every one of us from the lower middle class right on up to the 1 percent spent more on food to pay for those heirloom tomatoes, strawberries, corn, wheat, and chickens that cost $30—a never-gonna-happen *if*—we still wouldn't have heirloom flavor in the quantity we need. There isn't enough land. The population of North America has more than doubled since 1948, while vast tracts of the best farmland have been eaten up by strip malls, golf courses, factories, and houses. The more land we use to feed ourselves, the more we encroach on the dwindling parcels of nature that are still left. Yield may have its downsides, but the depressing truth is we need it. Until someone figures out a way to crack the existential tradeoff between quantity and quality, real flavor will remain expensive and only a few will be able to afford it.

For everyone else, pass the Doritos.

PART THREE

THE DELICIOUS CURE

EIGHT

The Tomato of Tomorrow

B Y THE spring of 2005, it had been ten years since Harry Klee left Monsanto to dedicate himself to cracking the mystery of delicious tomatoes. During that time he'd made the following pivotal, potentially world-changing, but undeniably eggheaded discovery: Tomatoes make phenylethanol from phenylacetaldehyde, which they make from phenethylamine, which—bear with me here—they make from phenylalanine.

There was only one chemical on this list anyone generally cared about: phenylethanol, which smells like rose-scented perfume and has long been considered one of the most important flavor compounds in tomatoes. No one, however, had ever bothered asking just how a tomato makes its much-loved rose note. To Klee, this was a crucial question. If he could figure out how a tomato manufactures its flavors, then he could find the genes that control flavor production, which meant he could breed for flavor. And that's just what Klee did. Like a Victorian explorer searching for the headwaters of the Nile, he followed a tomato's rose note all the way to its source: an enzyme that converted phenylalanine into phenethylamine and got the whole rose-scented ball rolling. At last, he had his hand on one of the levers that control tomato flavor. Getting to this point had taken a decade. It was a breakthrough. To Klee, at least.

Everyone else thought it was a headache. The big seed companies, including Klee's old employer, Monsanto, wanted nothing to do with the genetics of flavor, because the research was intricate and expensive, and the world seemed perfectly happy drowning their tomatoes in ranch dressing. They ignored him.

There was, however, one exception: a sprawling multinational called Syngenta Biotechnology. Among Syngenta's twenty thousand or so employees was a specialist in cell physiology named Steve Goff. A few years earlier, Goff led the team that won the race to sequence the genome of a domesticated plant—rice—and for this prestigious and celebrated feat *Scientific American* named him Research Leader of the Year. At Syngenta, Goff held the lofty position of senior fellow. He wasn't paid to develop new products or manage an existing one. He was paid to sit in his office and think deep thoughts.

Two such thoughts had been preoccupying Goff, and they both had to do with flavor. The first was dilution. Goff knew fruits and vegetables were getting blander, and he thought it might make good sense—good business sense—for Syngenta to focus on flavor. Harry Klee was just the guy to talk to.

The second deep thought was even deeper. Goff had begun wondering why people were attracted to certain flavors in the first place. His curiosity led him to the work of Fred Provenza. He was impressed by the elegance of Provenza's experimental models, but he was even more impressed by what those experiments suggested: that flavor and nutrition are intrinsically related for goats and sheep. Was there some way of proving that the same thing was true for humans?

Once again, Harry Klee was the person he needed to talk to. Not only did Klee know what flavor chemicals were important in a tomato, he also knew more than anyone else about how tomatoes make flavor. And that, for Steve Goff, was a critical question. Some aroma chemicals, he noticed, tended to pop up all over nature. Phenylethanol is a prime example. It's in tomatoes, grapes, kiwis, and apples, and it's also in roses, petunias, and scores of other flowers. Humans, furthermore, have an enduring love of the stuff. Phenylethanol is considered one of the most important compounds in the flavor industry—it's added to foods and soft drinks, perfumes, soaps and, of course, cigarettes. What made phenylethanol so special?

In April 2005, Steve Goff invited Harry Klee to Syngenta's head office in North Carolina to give a talk. In an auditorium filled with about forty people, Klee described the intricate metabolic process behind a tomato's rose note. While the flavor world was interested in the end product, an aromatic compound humans love, Goff saw it in a different light. As someone who studied the innermost workings of cells, his attention was caught by the chemical that started the process: phenylalanine.

For those not intimately versed in organic chemistry—I am a card-carrying member of this group—phenylalanine is an amino acid. Amino acids are what the body uses to make complex organic molecules, and their biological importance cannot be overstated. Amino acids are the building blocks of things like skin, hair, muscles, toenails, neurotransmitters, and even smell receptors. Phenylalanine, however, isn't an ordinary amino acid. It's essential. You must eat it, and if you do not, you will get sick and die. But even among the essential amino acids, phenylalanine is notable—to a guy like Steve Goff, at least. It's "expensive," metabolically speaking. It's a big, complex molecule with lots of molecular bonds that requires a lot of energy for an organism to make. Was it just a coincidence that a highly desirable amino acid was the building block for a highly desirable aroma? Or was this enticing rose-scented note a big neon sign saying, "Free phenylalanine!"?

For Klee, it was "a Homer Simpson moment." He'd known since his days as an undergraduate that phenylalanine was an essential amino acid. In all his work on tomatoes, he'd just never stopped to consider it.

After Klee's talk, the two men returned to Goff's office and, sitting across from one another, went through the list of what were considered the twenty most delicious aromas in tomatoes, compound by com-

pound. Beta-ionone, which smells floral and fruity, was made from a
carotenoid. Trans-2-heptenal, which smells "fatty, green, and pungent,"
was made from the omega-3 fat called alpha-linolenic acid, which is
used by the body to control inflammation, blood clotting, and build cell
membranes in the brain. As they knocked off every compound in the
top twenty, the pattern became undeniable: Each one was built from a
nutrient of life-and-death importance. Some came from essential fats,
some came from essential amino acids, and some came from carotenoids, which are necessary for sight and which the body uses to make
vitamin A. Over the course of one densely insightful afternoon, Klee
and Goff drew up some of the most compelling evidence yet for nutritional wisdom in humans: The flavor of tomatoes is inextricably linked
to nutrition. Of all the four hundred aromatic compounds in tomatoes,
there are twenty that seduce us into eating this luscious red fruit, and
every single one of them is made from things our bodies need. Their
findings would eventually be published in *Science*, considered by many
to be the most prestigious scientific journal in the world.*

THE TOMATO sweetness breakthrough came six years later when Linda
Bartoshuk experienced her third most intense eureka moment as a scientist. This is saying something. You probably haven't heard of Bartoshuk, but you have likely heard of her first most intense scientific
eureka moment: the discovery of "supertasters," which took place when
she noticed that some people were more sensitive than others to bit-

---

* Goff is now a research professor at the University of Arizona, where he received
one of the largest grants ever given in the life sciences from the National Science
Foundation. Following his work with Klee, he wrote that the enhancement of
food with natural and artificial flavors could "lead to confusion on the nutritional
value of these foods." He believes one of the reasons for the cultural enthusiasm
over new and trendy flavors is that the food people eat is not meeting their
nutritional needs.

ter, sweet, and salty tastes. Bartoshuk is a psychophysicist. She specializes in rating the intensity of experiences. The whole reason she joined Klee's research effort was to help get more accurate ratings on tomato experience. And that data proved to be so profound it caused the professional rater of experiences to experience an experience worthy of rating.

The experience was as follows: Bored in her office one afternoon, Bartoshuk decided to run some statistical correlations on tomato data, and she "almost fell off her chair." Some tomatoes, Bartoshuk had found, tasted sweeter than they should have.

There is nothing in and of itself unusual about a sweet tomato. But what made these sweet tomatoes special wasn't how much sugar they had—it was how little. A Matina tomato, for example, contained less sugar than a Yellow Jelly Bean tomato, and yet sensory panelists claimed it was sweeter. It was fully twice as sweet as the Jelly Bean. This didn't make sense.

And then, very abruptly, the lightbulb went on. Bartoshuk recognized an obscure phenomenon no one in her field had thought much about in years, one she herself had once called an "illusion." It is called "volatile-enhanced sweetness" and was discovered in the late '70s when scientists had shown that if you added a fruity aroma to sweetened water it tasted even sweeter. But the effect was tiny—so small no one, including Bartoshuk, took it seriously. And now here was volatile-enhanced sweetness raging in tomatoes. And it wasn't small. It was doubling—doubling!—sweetness. Using yet more statistical correlation, Bartoshuk discovered how tomatoes were pulling it off: "additivity." They stacked the decks. Whereas the scientists in their labs only ever examined one measly aroma at a time and found barely detectable results, tomatoes were piling on six at a time and achieving big-time sweetness. Bartoshuk was looking at scientific proof that tomatoes possessed seemingly magical powers to delight.

Klee was, by this point, two years into the world's most epic heirloom tomato phase, planting every kind of heirloom he could find on a research farm sixty miles north of Gainesville, and taking them back to his lab where a sensory panel would rate each tomato and an Agilent 6890 gas chromatograph fractioned out the aromatic compounds. The goal was to understand everything there is to know about how humans feel about tomatoes. Part of that required understanding tomatoes, which is why Klee was growing hundreds of heirlooms and running them through his GC. And part of that required understanding what it was about tomatoes that humans like.

Enter Bartoshuk, who put together sixty-three questions designed to extract laser-accurate tomato readings from sensory panelists. The questionnaire posed the standard-sounding queries about flavor intensity, sweetness, sourness, and so forth, rated on a scale of 1 to 9, but it also posed unusual questions such as, "Please type the strongest LIK-ING OF ANY KIND YOU'VE EXPERIENCED in the space below" and "Please type the strongest DISLIKING OF ANY KIND YOU'VE EX-PERIENCED in the space below." Panelists were to compare things like "the most intense annoyance you've experienced," "eating your favorite food," "the most amused you have ever been by an anecdote," and "the best tomato you've ever tasted," locating each on the spectrum rang-ing from extreme pain to extreme joy. The data set coming out of the Agilent 6890 would then be correlated with the data set coming out of humans, and simple statistics would reveal which chemicals inside a tomato turned people on and by how much.

Already, much had been discovered. For example, the role of umami in tomatoes, Klee says, is overblown. Women, furthermore, appear to be more moved than men by tomatoes. When asked to compare the "best tomato ever tasted" with "the strongest LIKING OF ANY KIND YOU'VE EXPERIENCED," women rate tomatoes higher than men do.

Perhaps most interesting is the unearthing of why foodies feel the way they do about tomatoes. To answer that question, you need to know what a foodie is in the first place. Is a foodie someone who tweets photos in real time of the sea urchin and frozen milk foam with cucumber-balls-and-dill granita he or she was just served by the most famous chef in the world? Is it someone who, even at an office Christmas party, while the music is pounding, will take a sip of red wine and say, "I'm getting passion fruit, caramelized rice cakes, and toasted Pop-Tarts"? Or is it Debbie? Bartoshuk and Klee came up with their own elegant definition. They defined a foodie as a person who rates his or her liking of food closer to the best life experience. In the hierarchy of things foodies like, food is way up there.

And foodies, they discovered, have a different relationship with tomatoes than do nonfoodies. Nonfoodies care only about sweetness. Nonfoodies sound like calorie zombies. Foodies, on the other hand, are turned on by aromatic compounds. "They like tomatoes that are more complicated," Klee says, "that are less sweet and have more sophisticated volatiles." Foodies sound suspiciously like Danish wine drinkers. Not only do they enjoy tomatoes more than do calorie zombies, they enjoy them for different reasons.

And now the curious case of volatile-enhanced sweetness. The result was a revelation: Tomatoes that taste better but have fewer calories. Bartoshuk and Klee had discovered one of the tricks nature uses to make food delicious.

It is quite a trick. It promises, if not to solve all the world's problems, then to help fix one of the biggest: food. For as long as we've been arguing over nutrients, we've had a pretty good idea of what we ought to be eating. As Michael Pollan puts it, eat foods your great-grandmother would recognize. The problem is that your great-grandmother wouldn't recognize the whole foods of today—not if she tasted them. That's the part Klee and Bartoshuk are fixing. After decades of mouth-numbing dilution, real food is getting its flavor back. They are creating deliciousness that's good for you, craveable foods that won't land you on the wrong end of the Yale Food Addiction Scale. Compound by compound, enzyme by enzyme, they have begun extinguishing the Dorito Effect.

One ripe and sultry Florida morning in June, Klee was standing in the air-conditioned refuge of his office, talking and slicing. He handed me a red disc of tomato along with a warning: "People always ask me, If you grew a Florida tomato in your backyard and picked it when it was ripe, how would it taste? You have your answer right there. It has no flavor." He was correct. The tomato was crunchy, the skin tough enough to suggest leather-making potential, there was acidity and minor sweetness, but just the vaguest whisper of tomatoness, a whisper that vanished so quickly you wonder if was just a voice in your head. Decade upon decade of plant dilution were all being expressed in this perfectly red, disease-resistant, easily shipped specimen. Bland.

There is much in the office of Harry Klee that inspires tomato pessimism. For instance, the action shot of a fruit picker dumping a basket of bright green apples into a truck. It could be titled "Apple Pickers at Harvest," but Klee points out the apples are tomatoes—unripe and hard as baseballs. There is a poster of beloved heirloom varieties. Klee pointed to the Red Currant tomato, which is the closest thing there is to the wild ancestor of modern tomatoes, and said, "It's not very good." He thinks even less of the Super Sweet 100—"the sweetness just dominates." (By his own metric, Klee sounds like a complexity-loving foodie.) Of the Green Zebra: "It tastes like you're eating an unripe tomato." A bookshelf within reach of his desk bulges with tedious-sounding tomes—*The Enzyme Reference. Plant Anatomy. Plant Physiology. Plant Diseases. Enzymatic Reaction Mechanisms.* But there, at the bottom right, is a grand and important reference work that promises the cure to all that research drudgery: *Jim Murray's Whisky Bible 2009.*

Klee is tall and thin with dark hair and a longish face, wears polo shirts and New Balance running shoes and has a deep but soft singsongy voice. He grew up in Beverly, Massachusetts, eating "unremarkable" tomatoes, and earned a bachelor of science degree in psychology at the University of Massachusetts, which brought him to the conclusion that "much of psychology is bullshit." A nonbullshit class called Drugs and Human Behavior, however, led to the PhD in biochemistry, which would eventually lead to a postdoc in microbiology, Monsanto, the slow-ripening tomato project, and eventually the flavor problem. If the last twenty years of Klee's life could be described as one man's wrestling match with tomato blandness, Klee may at last be within two or three moves of a choke hold.

Using years of data from the Agilent 6890 gas chromatograph correlated with data from humans, Klee has come up with a new and even better list of the twenty most important aromatic compounds in a tomato. There is much overlap with the old list—parasitic wasps will be pleased to learn cis-3-hexenol is on both. However, just like the list he used with Steve Goff back in 2005, every compound on the new list, including the ones that enhance sweetness, is built from an essential nutrient.

If a farmer can grow a tomato with these flavor compounds at optimal levels, it may not be the all-time best most blow-your-head-off awesome tomato ever picked, but it will be good. More important, breeders will have a fast and straightforward way of knowing if a new tomato tastes good. Instead of waiting till the fruit is ripe and paying thousands of dollars to perform a sensory evaluation, they'll just snip off a leaf and perform a quick genetic test. Any tomatoes not programmed to be delicious will be transferred to the compost heap, and the unnatural and pernicious disorder that is blandness will be slowly eradicated. One day, old-timers may tell their young grandchildren about the days when tomatoes had no flavor, and it's all thanks to Harry Klee.

I know what you may be thinking: Isn't Klee just creating a better dumbed-down tomato? A nutritionally diluted tomato with a handful of flavor compounds, after all, is still a nutritionally diluted tomato—a Dorito tomato. Not quite. In a tomato, it is impossible to boost flavor without simultaneously boosting nutrition. If Klee wants to enhance the sweetness of a tomato, he has to crank up the carotenoids. And even if such a tomato were marginally diluted, that is easily remedied by taking another bite. Instead of eating three tomatoes over the course of a week, eat four. If a tomato tastes good, the likelihood of that happening is good.

Klee isn't there yet. At this point he knows what flavors people like. What he hasn't figured out are all the genes that make those flavors. There are around twenty-five thousand to choose from, and the lab area outside Klee's office is a testament to the nature of this challenge. Tables are strewn with boxes of tomatoes, clamshell containers of tomatoes, bowls of tomatoes, loose tomatoes, and precious little seedlings under grow lights. On harvest day, paper plates covered in half-eaten slices of tomatoes are strewn in a way that looks like the remains of a "healthy" lunch for preschoolers. Over near the door sits the Agilent 6890, and, in front of it, long glass tubes loaded with diced tomato chunks that look like bruschetta for astronauts.

It has taken ten years for Klee and his team to locate about half the flavor genes. Finding the rest will take a few more. Yet even once he knows which genes he wants, Klee can't just manipulate the tomato genome, like a sound engineer operating a mixing board, to his desired flavor settings, because this would be a genetically engineered tomato like the one he made so long ago at Monsanto, which is to say a highly controversial and polarizing tomato that will soak up millions of dollars in testing and alienate vast swaths of the marketplace. So Klee will improve his tomatoes the old-fashioned way, through simple plant breeding.* He will introduce each trait one at a time and then back-breed and cross-breed generation after generation after generation until, at last, a genetic test will reveal a tomato that is just right. At that point, Klee is going to grow that tomato, pick the fruit, slice it, taste it, and then take a long holiday.

---

* I'll say it one more time for good measure. Despite the fact that Harry Klee used to work for Monsanto, his tomatoes are not GMO.

IN THE MEANTIME, Harry Klee has some outstanding tomatoes, and he would very much like you to taste them. He handed me a disc. Where the commercial tomato was crunchy, this one was soft and juicy. Where the commercial tomato was bland, this one gave off a plume of aroma that swirled like a dust devil in my nasal cavity. The flavor of a commercial tomato flashes and then plunges into nothingness. This one built, climaxed, and lingered, like a symphony that finishes with a single long note. The thing had flavor, as both the Agilent 6890 and the nasal cavities of 250 humans have attested. But as different as those two tomatoes are—one the embodiment of Big Food's dark tendencies, the other flavor's bright new hope—they share much in common. If tomatoes could talk, the bland tomato would turn to the flavorful tomato and say, "I am your father."

The tomato I had just tasted had nothing to do with "molecular breeding" and targeting specific genes. To make it, Harry Klee took the heirloom tomato that the sensory panels consistently rated the highest for flavor—a grape-shaped little number called a Maglia Rosa—and bred it with an ultra-high-performing flesh machine called FLA 8059.

He was hoping for a tomato that offered some of the benefits of each parent: hardiness, disease resistance, a thick skin for easier transportation, and a bountiful crop along with reasonable flavor. Klee wanted a compromise. It would be like crossing a stand-up comedian with an Olympic athlete and hoping for a daughter who's a pretty good athlete and tells pretty good jokes at dinner parties. Klee did not get what he hoped for. He got hardiness and disease resistance. He got yield—the plant returned 80 percent as much fruit as the best commercial varieties. And he got flavor so good it was statistically identical to its heirloom mother. Klee got an Olympic athlete that has them rolling in the aisles in Vegas. He named the new variety Garden Gem.

Harry Klee has delivered the best news in agriculture since the

Chicken of Tomorrow contest: Yield does not have to come at the expense of flavor. The tradeoff between quantity and quality, the two most defining traits in agriculture, is not a zero-sum game. You can have both.

For this, we can thank the nose. The aromatic compounds in a tomato—those elusive vapors that impart character and complexity and make it taste sweeter—are very cheap, metabolically speaking, for a tomato to make. Whereas sugar is measured in parts per thousand in a tomato, volatile compounds are measured in parts per billion. You don't need much. We can also thank the fact that humans are not calorie zombies, because every one of those compounds is made from a compound of life-and-death importance. If you think of a tomato's flavor as a big neon sign saying "free nutrients!" the problem is not that a modern high-yielding tomato can't afford the energy to power that sign. The problem is that the sign got unplugged. Tomatoes didn't get bland as a direct and unavoidable cost of crop size getting bigger. They got bland because in the race to breed big crops that were disease resistant, hardy, and made it to store shelves without getting bruised, flavor just got lost.

The question is, Now that flavor has been found, will Big Tomato care? This is the side of the bland tomato problem that no amount of science can fix. This is the human problem. And the data Klee has gathered so far is not encouraging. In what may be the most depressing symptom yet of the Dorito Effect, no one in the tomato world other than Harry Klee seems to care about flavor.

When Klee goes to tomato farmers' meetings and gives talks about how tomatoes can be made flavorful once again, he is met by a thousand-foot wall of polite indifference. The big growers are not evil. They are worried about the same thing they've always been worried about: this year's crop. They want it to be big and they don't want it to get wiped out by some insect or fungus. The last thing they think about is flavor because no one is paying so much as a penny extra for flavor. The growers sell to the wholesalers, who are also hooked on quantity because their customers, the supermarkets, sell tomatoes to consumers who think everything should cost 99¢ per pound and who've never known any-thing other than cardboard tomatoes.

In his dreams, Harry Klee can see his tomatoes on the shelves of su-permarkets, plucked by gleeful shoppers who delight in their wonderful taste. In his nightmares, they wind up in some seed bank, an interesting footnote in the history of dilution, ignored by a world in love with ranch dressing.

NINE

The Gospel According to
Real Flavor

I DECIDED TO do something. I was upset—outraged, actually. I felt disgusted, hurt, disrespected, pissed off, alarmed, baffled, depressed, and bewildered that industry doesn't care about real flavor. If tomato farmers won't grow Klee's excellent tomatoes, who will?

Me. I wanted to make an entire meal out of them. Actually, I had a better idea. I wanted a chef to make an entire meal out of them. I wanted to place Klee's tomatoes in the hands of a master because I wanted to know: Was extinguishing the Dorito Effect just a theoretical fantasy? Or could it really be done? I wanted to know if it actually was possible to base a food system on real flavor and produce ingredients whose deliciousness is imparted by nature, not laboratories. It was one thing to imagine a future where flavor matters, it was quite another to taste it.

I contacted Douglas Hayes, Buckeye master. I needed land and Hayes had it—forty-two acres in California's Napa Valley. Hayes knew nearly as much about growing heirloom tomatoes as he did about growing Buckeye chickens, and he was dying to give Klee's Garden Gems a try.

What I needed now was a chef, which isn't as simple as it might sound. A lot of chefs pay lip service to bucolic purity—they put sweet-breads or duck heart on the menu, and the waiter is sure to tell you about the foraged mustard greens—but they can't seem to kick the habit of blitzing everything in modern-day flavor solutions like soy sauce, panko, entire bulbs of garlic, bacon, and sugar. I didn't want faux-heirloom Dorito cuisine. I needed a chef who wasn't going to club these ingredients senseless. Hayes was in agreement. "We want a chef that will let the tomatoes shine," he said. Despite the Napa Valley's reputation as the holy land for farm-to-fork purity, Hayes believed there were no more than three chefs who "got" flavor, and foremost among them was Larry Forgione.

Larry Forgione is often referred to as "the godfather of American cuisine." In 1977, as the head chef at Regine's in New York City, he preempted the local/artisanal trend by a good quarter century when he began seeking out particularly good ingredients from small farm-ers who raised them particularly well. In the early '80s, a forager he worked with introduced him to fruit of "incredibly intense flavor" from northern Michigan, and the two formed a company to make and sell preserves. Today, he is the director of the Conservatory for American Food Studies at the Culinary Institute of America's Greystone campus, in St. Helena, California—just down the road from Hayes—where he teaches "the true meaning and essence of what is farm to table" and continues his decades-long crusade against flavor dilution. Hayes paid a visit to Forgione one afternoon to deliver some Buckeye eggs and asked him if he'd be interested in cooking this dinner we were scheming up. Forgione said yes. Then Forgione said, "What else do you want to serve besides tomatoes?"

The godfather of American cuisine asks a good question. And the answer goes well beyond tomatoes. There are nascent stirrings of a qual-ity movement in horticulture, a collection of scientists, breeders, and growers whose definition of "improvement" extends to more than just quantity. Over a period of months, I reached out to these people. I described my vision of a dinner built on real flavor and asked them to send ingredients. The criteria were as follows: The varieties had to be new and they had to have been bred for either flavor or nutrition—although, as you will see, you generally don't get one without the other—and they had to deliver a somewhat sizable crop. I wasn't looking for heirloom ingredients. But I wasn't out to set a new yield record, either. I wanted naturally flavorful ingredients that ordinary people can afford.

In South America, ordinary people can afford a noteworthy spud called phureja. The phureja is the second most popular variety of potato on that continent. They can be purchased fried on street corners in Bolivia and Colombia, and their reputation for fine flavor stretches back into folklore. Outside South America, their reputation for flavor is pretty much nonexistent because you can't buy phureja potatoes anywhere else. Well, almost anywhere else. A biochemist in Dundee, Scotland, by the name of Mark Taylor—he and Klee have met at meetings—has bred a Scottish phureja. It grows well during the long days of a far northern summer, roasts crispy on the outside and soft in the middle and, one presumes, goes superbly with haggis. Sensory panels laud this potato for its distinctive flavor and creamy texture, and it yields a mere 10 to 20 percent less than commercial varieties and is so packed with carotenoids it looks like an egg yolk with a suntan.

Taylor named the variety Mayan Gold, and short of smuggling a suitcase of Mayan Golds in from Britain, I discovered that it is simply not possible to eat one on North American soil because Taylor licensed the variety to a big North American potato company whose name he was forbidden to share with me. (I contacted every large potato company I could find and not one admitted to owning the Mayan Gold. Why remains a mystery.) Taylor suggested I contact Washington State University, a hotbed of potato progressivism, and there I found Chuck Brown, a breeder with a carotenoid grenade of his very own, a phureja-commercial cross that's so flavorful you can smell a faint nutty aroma when they're dropped into boiling water. The variety was, at that moment, languishing on a shelf in cold storage. Brown had offered it out to potato growers and not a single one wanted it. You're welcome to it, he said. I was also welcome to another potato, one that boasted some impressive health attributes and was enjoyed by sensory panels. These potatoes were purple and so packed with plant secondary compounds—anthocyanins, phenolics, carotenoids—that in two separate studies they had been found to lower blood pressure in adults with hypertension and improve inflammation markers in male college students. A few days later, a FedEx box was delivered to Douglas Hayes's farm in the Napa Valley, and a few days later Chuck Brown's unloved potatoes were nestled in earth.

The next box to arrive at Hayes's farm was filled with chicks—beige, fuzzy, and peeping. These were, in many ways, the very opposite of Hayes's beloved Buckeyes—bred by Hubbard Breeders, the world's third-largest broiler genetics corporation, a sprawling multinational responsible for inconceivable tonnage of blandness. Hubbard's broiler range includes the Hubbard Classic, which reaches 4.68 pounds at thirty-five days and offers "maximum throughput of broilers per site per year"; the Hubbard JV, which reaches 4.27 pounds at thirty-five days and "is ideal for the markets looking for the lowest carcass cost"; and the Hubbard H1, which reaches 4.55 pounds at thirty-five days and is "designed to produce the maximum saleable meat of any broiler breeder available on the market today."

There is more to Hubbard than blandness, however. The company also has a line of "slow growth" chickens that "respond to the demand for superior quality." (A tacit admission, it would seem, that broilers are no good.) These chickens are mostly bred for a French line of poultry called Label Rouge, which, you may recall from chapter 2, is the best chicken Chicken of Tomorrow award-winning Poultry Boy Paul Siegel has eaten since his mother's fricassee. The Label Rouge program was created in the early 1960s, when farmers in the Landes and Périgord regions of France noticed the same thing Julia Child did: that a modern chicken "needs strong dousings of herbs, wines, and spices to make it at all palatable." While Americans embraced teddy-bear stuffing, over on the other side of the Atlantic, the French government decided to do something about the emerging flavor problem by creating a brand that denoted chickens that actually taste like chicken. For a chicken to qualify, it must be raised partly outdoors, eat a particular diet, and be at least eighty-one days old when killed (some are over a hundred). Label Rouge chickens are regularly subjected to sensory panel evaluations and consistently top broilers in "flavor," "texture," and "appearance."

And so, if you have ever fantasized—as I do probably once a week— what might happen if the men and women trained by Paul Siegel set their minds to improving the way chickens taste instead of how efficiently they produce meat, that dream has been realized by a giant multinational. Hubbard Breeders has bred chickens that reach a flavor-tenderness equilibrium at twelve weeks—somewhat faster than a barred rock, but way, way slower than a Hubbard JV. Many of these chickens incorporate a breed known as a Transylvanian Naked Neck, which looks like a vulture and features thin skin that gets extra crispy in the oven. Label Rouge chickens boast less fat and water than industrial broilers, and they have more protein, minerals, and vitamins, with a respectable feed conversion ratio of 2.89. (A broiler clocks in around 1.6.) Last year, more than a billion dollars in Label Rouge chickens were eaten in France. Tomato growers of Florida take note: There is money in flavor. Ask Hubbard Breeders.

The University of Florida is banking on the future value flavor. In 2010, it created the Plant Innovation Program, a coordinated, multi-disciplinary assault on blandness in which forty scientists are applying Harry Klee and Linda Bartoshuk's handiwork to peaches, oranges, blueberries, basil, and strawberries.* A petunia expert named Dave Clark,

---

* A PhD student of Bartoshuk's named Jennifer Stamps, researching the effect of neurodegenerative diseases on flavor, put some U of F strawberries out at "daycare" for Alzheimer's patients one afternoon. "When you put fruit out from the grocery store," she told me, "it usually just sits there," in part because

who oversees the program, told me they've already discovered sweet-enhancing volatiles in strawberries and they're not the same ones as in tomatoes. Some oranges are similarly "too sweet for their sugar content," Bartoshuk told me. It is expected that blueberries and peaches will reveal their own mysteries. And the Plant Innovation Program will use this knowledge to make these fruits taste better, and people will buy them and eat them.

Klee's approach is being applied to another much-loved food at risk of ruination: chocolate. Ed Seguine, whom Klee calls the Robert Parker of chocolate, was until recently the chocolate research fellow at Mars Chocolate, a job he described to me as saving the world from a fate of bland chocolate. (He now consults for the Guittard Chocolate Company.) Already, there is chocolate that promises to do just that, and Seguine sent some to Larry Forgione at the CIA. Dave Clark promised blueberries, basil, and strawberries. It would all be prepared and eaten on the night of August 22.

RIGHT ABOUT THEN, the bad news started rolling in. On the afternoon of July 18, Douglas Hayes called to register a note of concern about his crop of Garden Gems. "They've just started setting fruit," Hayes said. Two days ago there had been the remains of a yellow blossom, now a tiny bulge of green flesh was pushing its way into the world. That bulge was three weeks late.

The seeds, which had come all the way from Gainesville, had gone in the ground May 9. By mid-July, the branches should have been drooping with green orbs waiting for their first blush of orange. But May was so cold and rainy Hayes had to dig the plants up and take them into the

---

Alzheimer's degrades the flavor-sensing system, causing many patients to become disinterested in food and lose weight. The flavorful strawberries, however, were "gone in a few minutes," and one patient began reminiscing about picking strawberries as a child in West Virginia.

greenhouse. June was worse. He was calling to give me the heads-up. The tomatoes were running late.

No problem, I said. There was a backup crop in Tennessee. Everything was on track for August 22.

The strawberries got nailed. More bad weather. Blueberries were a definite no. And basil was iffy—the stuff grew like a weed, but the intellectual-property folks at the University of Florida were uncomfortable with sending fresh cuttings that could, in theory, be propagated by an unscrupulous grower. Then there was a chicken issue. Hayes said they were growing too fast to be eighty-one-day birds, even though the breeder I'd sourced them from swore up and down they were "authentic" Label Rouge birdies. I contacted Hubbard directly and learned the truth: They were "medium growth" sixty-three-day chickens—still tortoises compared to modern broilers, but not true Label Rouge. More bait and switch. But not a total disaster. Hayes said he would grow them out past eighty days just the same.

And then the worst news possible: The backup Garden Gems in Tennessee had been victims of an infestation of fall armyworm. Finished. Kaput. The entire crop was ruined. The tomato that had inspired the whole dinner might not be at the dinner.

I began pestering Douglas Hayes for tomato updates. We'd been having daily conversations for some time—about the exact point at which chickens begin to take on flavor (week 9); the preponderance of bad sourdough in San Francisco (the best starter culture is available from a guy in Idaho named Ed Wood, Hayes says); and what rice to use for risotto (not arborio but carnaroli, according to Hayes). But now, like pubescent boys, we fixated on tomato anatomy: How hard is the biggest one? Is it longer than your pinkie?

Hayes said if they didn't get their first flush of orange a week before the dinner, they wouldn't be ready. That day was August 15. The morning of August 15, I phoned Douglas Hayes.

"What color are the tomatoes?"

"Green."

"Is there any sign of orange?"

"No."

WITH TWO DAYS to go before the dinner, I found myself on a test farm in California's Santa Clara Valley, in a state that might best be described as a controlled yet foggy panic. A hot wind was blowing. I was bent over a table covered in light green foliage, which I petted like a cat, and was struggling with McRegret.

Forty-five minutes earlier, I had pulled off the interstate in Palo Alto to satisfy the need state of extreme hunger. I pulled in to a strip mall and grabbed a slice of pizza, a standard North American corruption of too much crust and industrial mozzarella. It tasted good going down, but the megaload of carbs and fat induced negative post-ingestive feedback and I pulled back onto I-101 feeling bloated, exhausted, and mentally fogged.

The plant I was now petting was sorrel, which will be familiar to home gardeners, herbcentric foodies, scurvy-ridden eighteenth-century sailors foraging on Juan Fernández island, and almost no one else. David Griffin, who is the CEO of the Shamrock Seed Company, is working to change that. Griffin believes consumers will take to sorrel because of the way it tastes—fruity and nearly sharp, but in a pleasant sort of way. The leaves, unfortunately, have a tendency to get progressively woodier and bland as the plant matures. So Shamrock is breeding a new sorrel with a fatter delectability curve.

Sorrel is one among numerous attempts by Griffin to add flavor to the vegetable life of America. Most have been failures. There was Gemberg, a lettuce with "phenomenal texture and flavor" that isn't catching on. There was the Chinese leaf lettuce that kept getting overrun by mildew. And there were carrots. Way back, at the start of his career, Griffin had a meeting with a California carrot mogul, the man who, years earlier, came up with the idea of carving garbage "cull" carrots into cute little "baby" carrots, thereby revolutionizing the industry (and putting some people who sold actual nongarbage carrots out of business). The younger, naïve, idealistic David Griffin thought (and still thinks) flavor is a big reason why humans put vegetables in their mouths and tried to sell the carrot mogul on this radical idea. The carrot mogul was all for it. "How many dollars should we put toward carrot flavor?" Griffin asked.

"Spend all the dollars you want," the carrot mogul replied. Griffin was elated. Then he heard the conditions. The flavorful carrots had to cost the same, had to yield the same, and had to have the identical long and tapering shape as bland carrots so they would fit perfectly inside the carrot-carving machines. Flavor was all well and good so long as it was free. The conversation taught Griffin everything he would ever need to know about how the produce industry feels about flavor. For reasons he is unable to fully articulate, he persisted.

Griffin puts his success rate at about 3 percent, and so far the signs are good that sorrel will join this elite group. It is currently featured in several "lemony blend" salad mixes sold at various supermarkets. Griffin's company now moves 2,000 pounds of sorrel seed a year, which works out to 130,000 pounds of sorrel going into the mouths and bodies of humans every month. That is good. But it is not as good as arugula, an inspirational defy-the-odds story of nutritional wisdom.

When talk of arugula first started back in the early '90s, the big players in the lettuce industry all had a good laugh. The people who grew and sold salad considered it nothing more than a vehicle for dressing. And nothing carried a dressing like crunchy, high-yielding iceberg lettuce, the vegetable that tastes like frozen water. Millions of pounds of lettuce came out of the Salinas Valley back then, and three-quarters of it was iceberg. None of it was arugula. When it came to flavor, the industry lived by the following maxim: "Bland is what people want." Arugula was many things—dark green, peppery, bitter, loaded with plant compounds—but it sure wasn't bland.

Amid all this resistance, a few companies took the imprudent gamble of including leaves of arugula in their salad mixes. It wasn't even the hard stuff—the "wild" arugula they eat over in France and Italy. This was a "cultivated" arugula that has softer leaves and a milder flavor. Consumers loved it. The response was such that by the mid-2000s, arugula wasn't only in mixes anymore. It got its own bag. Soon cultivated arugula was out, and Americans, acting like Utah goats, switched to wild arugula. They liked the bitter flavor. They loved the peppery intensity. Americans now eat twenty million pounds of arugula a year. Bland, it seems, is not what people wanted.

If, back in 1992, all those bland-loving produce barons got in a time machine and visited the grocery store aisles of today, they might think they were in Europe. Leaf lettuces are crushing it. Chopped iceberg is a loss leader, the cost of getting shelf space for the exotic blends, which is where the real money is. Five years ago, kale was openly mocked in the salad industry as "cattle food." Now kale, chard, and beet greens are the industry's fastest-growing segment. The arugula story isn't finished. Griffin has bred a new variety that pushes the flavor past wild all the way to wasabi. It is primo stuff, sufficiently potent enough to make you aware of sinus passages you didn't know you had.

After my introduction to Shamrock's sorrel, we went on a tour of Shamrock's test farm, sampling genetic permutations designed to create real flavor that might squeak into the 3 percent. We ate mustard greens from Ethiopia with a soft leaf and gentle but pronounced flavor that Griffin's breeder named after the country's Amhara tribe. We ate a peppery watercress called peppercress, a creamy Japanese salad green called tatsoi, a clover that tastes like baby corn and radicchio, arugula flowers, a variety of dill that has notes of licorice, and three kinds of minikale. When we were finished, a curious thing happened: The McRegret went away. The bloat dissipated, the mental fog lifted, and for the first time since lunch, I could think straight.

And then we really got down to leaf eating. I wasn't here for sorrel, peppercress, wasabi, arugula, or tatsoi. I was here to sample some of the most exciting leaves Griffin has ever cultivated. They are the brainchild of a plant scientist at Rutgers University named Ilya Raskin, who set out to grow a plant with enhanced nutrition. It had to grow fast (tree fruit was out; so were tomatoes); it had to be nutritionally benign (neither healthy nor unhealthy); and it had to be a food people ate a lot of. This left a single, obvious choice: lettuce. Raskin took lettuce leaves and broke them down into tiny cell clumps and grew them in petri dishes, a process that encourages new and distinct genetic lines. When the clumps got bigger, Raskin tested them for a family of health-promoting secondary compounds known as polyphenols, and the clumps that tested highest were planted in soil. After three generations, Raskin had a new and profoundly different lettuce, one with intensely purple leaves and two and a half times the polyphenol density of blueberries. Raskin called it Rutgers Scarlet Lettuce, and in March 2013 the Shamrock Seed Company purchased the rights. The leaves are soft and silky with, as Raskin puts it, "a pleasant bitterness, somewhat like radicchio but milder." He may have been after nutrition, but he still wound up with flavor.

The Dorito Effect has a new enemy in Ilya Raskin. Just as Harry Klee is fixing what happened to flavor, Ilya Raskin is fixing what happened to nutrition. One day, it might be possible to purchase a tomato—or a strawberry, lemon, kumquat, avocado, mango, you name it—bred for both. One day, farmers will be able to grow produce bred to please growers and eaters alike, and agricultural "improvement" will finally live up to its billing.

Whether you or I or Ilya Raskin will be able to buy Rutgers Scarlet Lettuce at the supermarket remains a big if. Griffin is already getting guff from the industry. The morning we visited, a buyer from one of the country's largest lettuce companies said, "If it's not 99 cents, I'm not interested." (It will cost around $3.60 per bag, 20 percent more than standard bagged lettuce.) Another buyer told him consumers do not care about nutrition. Others worry that Rutgers Scarlet Lettuce will cannibalize sales of other lettuces that generate a higher profit margin. Worst of all, pennies from every bag sold will go toward funding further research. Bland may not be what the people want, but the industry sure loves it.

GRIFFIN DID have one piece of truly good news, which I sorely needed considering the dinner was less than forty-eight hours away. In his office was a small box, inside the box was a small cooler, and inside the cooler were nine pounds of ripe Garden Gem tomatoes. They'd been delivered a few hours earlier by a UPS truck, which received them, if you follow the metabolic chain all the way back, from a UPS store in Toronto. I hadn't seen the box in twenty-four hours. I had every reason to think it would never make it to California, but here it was. I wanted to hug it.

When Hayes dropped the troubling news about his unripe Garden Gems, desperation set in. Months earlier, Klee had sent me a packet of Garden Gem seeds, which I had distributed to various friends who were, like me, under the heirloom tomato spell. As the disasters in California and Tennessee unfolded, I called them up, beginning each conversation with, "Hey, remember those seeds I gave you?" A criminal prosecutor I know picked two ripe tomatoes just yesterday. He ate them. A farmer two hours west of me planted his tomatoes late and the fruit hadn't even set yet. A farmer an hour and a half north had four plants, each branch heavy with fruit, all of it green. I checked the weather in Napa every half hour, wincing at clouds, nearly weeping at predictions of sunshine. I imagined beams of solar energy driving photosynthesis, coaxing these stubborn plants into a ripening miracle.

The farmer who had four plants laden with green tomatoes e-mailed me one morning—still green. But he'd given some seeds to a farmer down the road, and that guy's tomatoes were red. Rapture. Ecstasy. A friend in the city had a single Garden Gem plant that was producing like mad, and I was welcome to whatever I could pick. In the end, I managed to cobble together a harvest from two sources as opposite as this tomato's curious parentage: seven pounds from an Amish farmer with nine kids and two pounds from a neuroscientist who studies lipid signaling in the brains of humans and rats.

There remained a single impediment, one even more massive, complex, unpredictable, indifferent, and prone to ill mood than the weather: the government. It is all well and good to have nine pounds of Garden Gem tomatoes in Toronto. But getting them over the border into the United States and all the way to the Napa Valley—and doing so in a way that does not compromise flavor—is another matter altogether.

For example, if you phone the USDA Animal and Plant Health Inspection Service's Plant Protection and Quarantine Import Permit Service and press 1 for "questions pertaining to importation of fruits or vegetables," you will be told that as a noncommercial grower of tomatoes, it isn't possible to export tomatoes to the United States. If you phone back two minutes later and this time press 9 to speak to a plant protection and quarantine import customer service representative, you will get a different version of the supposed truth. All you need, you will be told, is a "certificate of origin," a "phytosanitary certificate," and a commercial invoice, but even then it could still get turned back by a customs officer for no particular reason. If you then call the Canadian Food Inspection Agency, the outfit on this side of the border that issues phytosanitary certificates, you will be informed that the USDA APHIS guy has it wrong, that a phytosanitary certificate is *not* required if the tomatoes are of Canadian origin—and the CFIA won't issue one no matter how much you're willing to pay—and that a certificate of origin is nothing other than a signed form in which the "exporter" (in this case, me) attests that the tomatoes are of Canadian origin. If, at this point, you ask the CFIA guy in a hey-I'm-just-curious tone of voice what's to stop a person from "attesting" to the Canadianness of an imported Honduran tomato because no one would ever know any different, you will be met by an uncomfortable silence. In time, you discover there are two certificates of origin templates, one in the *Canadian Border Agricultural Clearance Manual* and another known as a NAFTA Certificate of Origin, and you spend two days wondering which one is preferable. You decide: to heck with it, use both. You spend the better part of a week madly Googling to find out what the difference is between a commercial invoice and a customs declaration until Yashar at the UPS Store tells you, "Don't worry, my friend, they are the same thing." You discover, with only days to spare, that you need an FDA prior notification number, just as you discover that 0702.00 is a general Schedule B number, so you debate using 0702.00.0025, which designates "Roma" tomatoes, and 0702.00.0015, which designates "cherry" tomatoes, and ultimately decide that the Garden Gem is technically neither and go with 0702.00.0035, which is the specifically nonspecific Schedule B number for "other" tomatoes. With all the forms corralled in a single precious envelope that you clutch to your chest lest there is a sudden gust of wind, you carry said envelope along with a cooler full of Garden Gems tenderly packed in newspaper, foam, and bubble wrap into the UPS Store on Pleasant Boulevard, where Yashar gently tells you he cannot guarantee the temperature in the cargo hold won't go below freezing though he's pretty sure it won't. Knowing what you know about the deleterious effects of cold temperatures on a tomato's flavor compounds and the random temperament of customs officers, you find yourself locked in a state of gasping turmoil until the next morning, when your flight touches down in San Francisco and you turn on your cell phone and receive the greatest e-mail in the history of the Internet. It is from the Shamrock Seed Company and reads as follows: "The tomatoes have been delivered!" A tear in your eye, your thoughts turn to the goodly customs officer, standing amid cages of exotic animals and boxes of rotting bananas with no Schedule B number. She examines your paperwork and, thrilled by its superb state of completeness, waves the tomatoes through. Later that day, at the offices of the Shamrock Seed Company in Salinas, California, you open the lid of the cooler Yashar taped shut a lifetime ago, inhale the intact tomatoey plume, and in no way regret the $405 cost of shipping.

I WAS, at last, northbound with tomatoes. I passed pizza-tragic Palo Alto, Redwood City, and San Mateo, and crossed the Bay Bridge to Oakland for lunch at a restaurant called Oliveto, whose proprietor, Bob Klein, handed me a four-pound bag of Hollis wheat kernels, a variety released by Washington State University as a low-input wheat (it doesn't need much fertilizer) that yields decently and, most important, is prized for its nutty top notes. (Note: Oliveto's stone-ground Hollis wheat penne Bolognese is the greatest restaurant pasta I have eaten outside Italy.) I continued north to a fruit and vegetable stand on a country road outside the town of Sebastopol, where from July through September Shelton's Market Garden sells a variety of strawberry unlike anything you've ever put in your mouth called Mara des Bois. The berries achieve the radiant hue of red that sports cars and electric guitars aspire to, and each is so tender you don't pick a Mara des Bois so much as snip the stem and airlift it to safety. In the hand, it is a perfumey disturbance. The waft of aroma hits you even when the berry is still at belt level, and when you take your first bite your eyes bulge and you laugh. (This also happens on bites two through eight.) It is a controversial aroma. Some people question whether Mara des Bois strawberries taste like strawberries at all because there is an undeniable grape note high in the mix, the same grapey note—the plant compound called methyl anthranilate—that's in concord grapes as well as drinks like AriZona Grapeade. The truth is the grape note is there in wild strawberries, too. It just got lost somewhere on the road to yield.

I bought a flat of Mara des Bois, braced it on the backseat with my luggage, and headed east toward Napa. Aroma was curling off the strawberries like smoke from a tire fire. Whenever I came to a red light, it rolled into the front and soaked me, eliciting expletive-laden astonishment. When I pulled away, it receded. I got to Hayes's place around dinnertime. He lifted the strawberries out of the back, popped one in his mouth, and said, "These are the real deal."

WHEN THE Culinary Institute of America took over the old Greystone Cellars winery in 1993, a building that had been devoted to the production of wine for more than a century switched to the production of chefs. The Conservatory occupies the old gatehouse, which sits behind a stone fence, swaddled in foliage like some fairy tale cottage where a little boy finds a kindly wizard who serves him a bowl of enchanted soup. The truth is not far off. During fifteen-week sessions at Greystone, Forgione steeps his students in the lost flavor of the ancients: heirloom tomato cultivation, cured sausages from pigs that eat acorns, Buckeye chickens and eggs.

On the evening of August 22, guests parked their cars in the lot—fairy tales set in California feature plenty of parking—and Forgione started them off with a cocktail that no organism, human or otherwise, had ever tasted before. The cocktail was made from honey-roasted oranges, elderflower liqueur, egg whites, gin, and "basil #1," a University of Florida cross that is plentiful and disease resistant and tastes profoundly like what it is: the offspring of regular basil mated to lemon basil. As Forgione's students applied their sharp blades to Garden Gem tomatoes and gently tore leaves of Rutgers Scarlet Lettuce, out in the dining room the guests sipped and chatted in retronasal fulfillment.

The guest list consisted of anyone I could think of who might be interested in the intersection of flavor and nutrition. I had put the invite out months earlier, thinking I might convince two, maybe three people to travel to Northern California and eat, among other things, Chuck Brown's purple potatoes that bring down blood pressure. But it turned out that pairing the godfather of American cuisine with the cutting edge in flavor cultivation is quite the draw. There were sixteen guests in all, and they included Jay Olshansky, a MacArthur-funded longevity researcher at the University of Illinois with whom I'd had a three-year e-mail conversation mainly about steak but also about nutrition and living longer; Mitchell Davis, the executive vice president of the James Beard Foundation; Ed Seguine, the chocolate research fellow at Mars; Douglas Hayes; Harry Klee; Dave Clark, who not only brought the basil but also bred it; and David Griffin.

Over by the door, Ed Seguine was sipping a virgin basil #1 cocktail and could be heard to say, "The world will need a million more metric tons of chocolate in a decade." And it will need a million more the decade after that, and a million more the decade after that. Whether the world gets that much chocolate, and what that chocolate will be like, is the problem Ed Seguine has dedicated the closing chapters of his career to solving. Many cocoa farmers are turning to more profitable crops, like corn or rubber. They have mouths to feed. Some don't have a choice, because diseases with names like witches' broom and frosty pod are laying waste to untold cacao acreage. (Brazil used to be the world's third-largest exporter of chocolate. Now it's a net importer.) China's expanding middle class, meanwhile, is developing a taste for chocolate. So is India's. So is Brazil's. This mounting need state, furthermore, appears to be based, at least in part, on nutritional wisdom—chocolate's apparent healthful attributes include improved blood flow, cardiovascular health, and lower blood pressure, as well as anti-obesity and anti-diabetic properties. (The effects are strongest with dark chocolate, which is richer in secondary compounds and is what chocolate aficionados prefer, even though it's bitter.) As world production of chocolate appears set to decline, consumption may be set to spike.

The result will be very expensive chocolate. Or possibly very bad chocolate because chocolate producers have set their sights on yield. A variety is being grown in Ecuador that produces more than sixteen thousand pounds per acre—nearly ten times more than normal. There is one problem: "It tastes like acidic dirt," Seguine says. And the only way to make it palatable is to process it in such a way that removes the beneficial plant compounds. Chocolate, it seems, is on a collision course with the Dorito Effect.

And on that note, the godfather of American cuisine asked that we please take our seats.

Tasting of Buckeye and Hubbard Chickens, part 1
Chicken crudo with wasabi arugula
Twelve-hour chicken broth

* * *

Salad with Hollis wheat berries, Rutgers Scarlet Lettuce,
extra virgin canola oil

* * *

Quick-stewed Garden Gem tomatoes with homemade fresh
garaganelli pasta

* * *

Tasting of Buckeye and Hubbard Chickens, part 2
Roast chicken
Poached chicken with mashed and roast orange and purple
Chuck Brown potatoes, wilted Amara mustard leaves

* * *

Napoleon of CATIE R-4 and R-6 blend chocolate & Mara
des Bois strawberries

The Buckeye took gold in the chicken flavor Olympics that night. Not a surprise, exactly, considering that at eighteen weeks, the Buckeyes were older and had captured more pasture with their beaks. The Hubbard birds, which Hayes had taken all the way to eighty-four days, were bigger, milder, and slightly more tender. But they were by no stretch weak. And that is the point. If North American consumers ever begin to feel about the flavor of real chicken the way they began to feel about arugula in the late '90s, it would be very difficult to satisfy that need state with Buckeyes and barred rocks and other chickens of times past. There aren't enough of them. And the genetics, which for decades have been in the hands of well-meaning but often untrained backyard enthusiasts, are a mess. It took years for Hayes to breed his Buckeyes into shape. He sells them to locals for $8 a pound. He can't supply supermarkets, and even if he could they would balk at the price. Hubbard, on the other hand, can. In France, Hubbard produces Label Rouge chicks raised on farms that look and smell like farms as we like to imagine them, and people walk into supermarkets and buy them. Label Rouge chickens cost about double what broilers cost, and yet they command more than half the market for whole chickens. The takeaway is as follows: Big Food can be part of a better flavor solution. If there's money in real flavor, they will give people real flavor.

So attention, Big Food—and Little Food, and Medium Food—the flavor is there if you want it. It's in the Rutgers Scarlet Lettuce, which tastes "amazing," according to Mitchell Davis of the James Beard Foundation, who was "struck by the beautiful bitter flavor." Flavor is in the purple potatoes whose mash was "silky, sensual, and buttery" and in the orange potatoes, which were brighter than the flashlight I brought to summer camp in 1983 and tasted "awesome" according to Dave Clark. The star of the night may well have been Harry Klee's Garden Gem tomatoes, which collapsed into a sauce whose depth and concentration reminded one guest of her grandmother and moved Ed Seguine to remark he needed one of those tiny spoons like they hand out in ice cream shops so he could get every last drop.

At about the point when teeth were sinking into carotenoid-rich chicken skin, I began wondering, once again, why it was that when the

French noticed chicken getting blander they did something about it. Strange how the continent that embraced teddy-bear-stuffing chicken began an affair with arugula, and more recently went cuckoo for kale. Who, exactly, is to blame for the crisis in flavor? Is it the breeders and growers, with their perverse fixation on yield? Is it the supermarkets, which treat food as "inventory" and don't have so much as a single box to tick on their order sheets for flavor? Maybe it's all the flavor companies' fault, because it's their sprinklings of chemical lies that empower all this terrible food. Then again, it would all be nothing without the consumers, the horde of shopping cart pushers that thinks anything that costs more than 99¢ is a ripoff.

If the dinner proved one thing, it is this: There is no excuse anymore. We can begin growing all the food we need, and it can be flavorful and fulfilling in the way the human animal desires it to be. Technology got us into this mess, and technology can get us out. The rest is up to us. If consumers demand real flavor, and if they pay a little bit more for it, then real flavor is what they will get. It's already happened with wine. It's already happened with craft beer. Let's make it happen with food.

As Forgione and his servers began staging dessert, the plainspoken and grandfatherly Ed Seguine, who has loved chocolate for sixty-four years and worked with it professionally for thirty, got up to say a few words. "This chocolate," he began, "was an effort to breed for frosty pod resistance." Not the most poetic introduction a dessert has ever received, but that was the point. The chocolate was grown in Costa Rica by an agricultural research organization called CATIE, and no one other than Ed Seguine had ever tasted this particular blend. CATIE also bred for productivity—they wanted a yield two to four times normal. Seguine's intensity began to build. "And for the first time in three thousand years of chocolate history that we know of," he said, "they decided to include flavor as a criteria." The result is four new varieties of cacao trees that meet all three specs. They feature substantial resistance to frosty pod, the yield is five times greater than average, and, most important to every

ving food for better flavor should be a "food movement." "Less is
re," he said, and went so far as to draw a graph illustrating how plea-
e diminishes as quantity increases. "If every portion is small and in-
sely flavored," he explained, "then the entire meal stays at the place
greatest enjoyment."

And with that, everyone walked to the parking lot, got in their cars,
d drove off into the night. In eight hours, it would be time for break-
st.

OUGLAS HAYES'S Garden Gems held out for another two weeks. One
ool Napa Valley morning, Hayes went outside and found they had
aken on the first blush of orange. In a few days, he was eating Garden
Gems. "It's a damn good tomato," he told me. "I'm very pleased I've got
hem in my garden." One afternoon, Hayes plucked off several pounds
and drove them to Oakland, to an Italian restaurant called Dopo where
he eats once a week with his ninety-nine-year-old mother. The restaurant's chef, Jon Smulewitz, peeked in the box and said, "Look at those
beauties." He reached in and grabbed one, bit into it, and announced,
"There's only one word that describes the way these tomatoes taste.
Delicious."

person seated in the Conservatory and every huma   gro
loves chocolate—they have flavor. In 2009, two of   mo
were entered into Cocoa of Excellence awards at th   su
lat in Paris. Both were winners. CATIE R-4 was rec   te
cocoa, sweet, fruity, and floral notes, and CATIE R   of
for its nutty and woody notes with undertones of
chocolate—very typical, Seguine says, of Central Am   an
As a sales brochure might put it: Five times the yield!   fa

Seguine passed around little tasting squares. Choco
a food so much as a "journey" and must be experience
or an afternoon at the spa. "Put it in your mouth and ru   r
of your mouth with your tongue, and then bring in smal   c
bring the volatiles to the retronasal area." Air was sipped.
conveyed to the retronasal area. Receptors were stimulate
quenched, truly and lastingly, by reward. Unlike an afternc
the high-yielding frosty pod–resistant CATIE chocolate
round of applause.

Two of the people I most wished could attend that din
Fred Provenza was at a wedding in upstate New York. Maria
was stuck at the office. So I ended the night by asking the qu
might ask: How did everyone feel? We were well into the pos
feedback phase. Food was in the gut now, triggering sensors a
ing hormones. Feelings were being felt.

"I feel very satisfied," David Griffin said. "Sustained."

Mitchell Davis felt "good." He elaborated: "You're not sup
leave a meal feeling sick, and I consider it a fault of the chef if
Harry Klee was glowing, but this had a great deal to do with
that Ed Seguine, who possesses the most discerning palate Klee h
come across, loved his tomatoes. Amy Myrdal Miller, who is the
nior director of culinary nutrition, said, "I feel happy and lig
re to eat a meal where you're mindful the whole time."

The longevity expert Jay Olshansky declared this whole busine

Appendix:
How to Live Long and Eat Flavorfully

Eat Real Flavor

In nature, flavor never appears without nutrition. No morsel of food should pass your lips before you have asked the following question: Where did the flavor come from? If it came from the plant or animal you're eating, keep eating. If it was applied by a human with a PhD in chemistry, put it down.

Eat Like a Utah Goat

Eat a variety of real foods, including things you think you might hate, like broccoli, collard greens, liver, and mackerel. (Trust me: All are delicious.) Recognize that your palate is a growing, living thing. It can and it will change. What you liked when you were nine years old is not what you will like as an adult. Nibble new foods. Try something ten times before you know for sure you don't like it. Above all, eat foods you find deeply satisfying.

Flavor Starts in the Womb

Research shows that for infants (and cattle, too) the lasting effects of flavor start before birth. Babies are exposed to flavors via amniotic fluid and breast milk, and moms who eat healthy food beget kids who are less picky. Moms who eat junk food, on the other hand, tend to have kids who like junk food. The palate is a lifelong investment. Treat it properly.

Eat *For* Flavor

Eat the best-tasting real food you can find. If you think it's expensive, remember that it's going to a very important place: your body. Seek out carrots that taste sweet and carroty. Find peaches that are juicy, tender, and peachy. Choose varieties of lettuce that actually have flavor. (General rule: The darker the leaf, the stronger the flavor.) Try the branded tomatoes that cost more—they usually taste better—and opt for "virgin" cold-pressed oils over refined ones. (Though be aware that virgin oils don't always perform well during sautéing or frying.) Visit farmers' markets to find the farmer growing the best-tasting food. Try out the produce at more expensive supermarkets to see if it's better. And if the food you bought tastes like cardboard, complain. Complain to your spouse, complain to the waiter, complain to the chef, or complain to the person or supermarket you bought it from. No one will know you care if you don't say anything. Finally, sit back and think of all the money you'll save on ranch dressing, ketchup, and whipped cream.

Eat Meat from Pastured Animals

It costs more, but it's healthier and you will eat less of it. Choose grass-fed beef that's at least twenty-two months old—not the ultralean cheap stuff—and pasture-raised pork. Buy cheaper braising cuts that will perfume dishes and nourish your soul. The chicken situation is dire. Most "pasture-raised" birds are modern broilers that just happen to be squatting on blades of grass as they stuff their gizzards with corn and soybeans, and their appetites and growth rates are such that they can't consume enough pasture for it to make much more than a token contribution to their flavor or nutrient profile. Organic chicken is often worse—just a high-yield broiler that's fattened, indoors, on organic feed. Comb farmers' markets and specialty food shops for chickens no younger than nine weeks—and preferably twelve to eighteen—that have access to pasture in warmer months and are fed green feed when it's too cold to go outside. (If you can get your hands on a bird older than eighteen weeks, make chicken and dumplings, but have some tissue on hand to sop up the tears of joy.) Warning: You can't tell a good chicken by the color of its skin—billions of commercial broilers are fed pigments to make them look "free range," and some very fine heirloom varieties, such as the Houdan, Dorking, and Australorp (and certain Label Rouge hybrids), will have white skin no matter how much carotenoids they peck.* A good chicken appears less squat and obese than a broiler, and has longish legs and a diminutive breast that harks back to a more innocent era. Finding such a chicken isn't easy, but it is worth it.

Avoid Synthetic Flavor Technology

A single can of Coke, bag of chips, squeezable yogurt tube, or fast-food meal will not reprogram your palate and doom you to a future of extreme obesity. But each time you consume human-made flavors, you're tricking your brain (or, worse, the brain of a child). The more you do it, the greater the consequences. The less you do it, the less you'll like food like that in the first place. Read the ingredients.

THE FOLLOWING WORDS indicate the presence of chemicals that fool your nose:

---

* White skin is caused by an enzyme that cleaves carotenoids in half. The very same enzyme is found in tomatoes, as well as some wild salmon from the West Coast that have uncharacteristically white flesh and are considered by many to taste better.

natural flavor(s)
natural flavoring(s)
artificial flavor(s)
flavoring
flavor

The following words indicate the presence of chemicals that fool your tongue:

monosodium glutamate
MSG
disodium guanlyate
disodium inosinate
torula yeast
yeast extract
hydrolyzed protein
autolyzed yeast
saccharin (Sweet Twin, Sweet'N Low, Necta Sweet)
aspartame (NutraSweet, Equal, Sugar Twin)
acesulfame potassium (Ace-K, Sunett, Sweet One)
sucralose (Splenda)
neotame (Newtame)
advantame
stevia

Avoid Restaurants That Use Synthetic Flavorings

I was able to find just a single chain restaurant that publishes a list of ingredients: McDonald's. Within its 37 pages, synthetic flavoring appears, by my count, more than 220 times (not including yeast extract or other umami triggers). At least McDonald's is being up front. Other chain restaurants—even family-friendly ones with comfy booths and waitstaff that seem to genuinely like you—follow the same model: premade industrially flavored food that's assembled and rethermalized on premises. So do a lot of bars, cafeterias, taverns, diners, and pubs. (Ever wondered why all buffalo wing sauces, chicken nuggets, and Caesar salads taste strangely alike?) As your palate becomes tuned to real flavor, you will find the food served in these establishments to be over-seasoned, calorie-laden pap. Eat in restaurants that employ actual chefs who cook using dead plants and animals, not low-paid laborers who pour sauces out of plastic buckets and flavor powders out of bags. Patronize restaurants that honor the flavor of the ingredients they use and don't mask them in seasoning.

Organic May or May Not Save You

Since organic fruits and vegetables are less protected by pesticides and fungicides from natural enemies, they have to work harder to protect themselves by producing plant secondary compounds.

In theory, they should be more flavorful. But a lot of the time they're not because many "industrial organic" farms grow modern varieties that aren't capable of being delicious. An "organic" label is no guarantee that food will taste better or be better for you. The true test of quality is the way food tastes.

Eat Herbs and Spices

They are good for you. They make food taste better. But use them to complement the flavor of food, not to cover up underlying blandness.

Don't Pop Vitamin Pills

There is no scientific evidence that vitamin supplements improve health and every reason to believe they empower a bland, high-calorie diet. Even worse, if your body is in need of vitamins, popping a tasteless pill could set up an unintended flavor preference just as Fred Provenza did with his coconut-loving sheep. Eat real food whose flavor tells the gripping story of its nutrition.

Eat Dark Chocolate and Drink Wine

And craft beer, too. These foods don't prevent or cure disease all on their own, but they are the mark of palates in tune with good food. Think of them as gateway foods to a healthier palate.

Give a Child an Amazing Piece of Fruit

The joy adults take in offering children sweet treats is so universal that it's probably a hardwired instinct. So make that treat a stellar peach, wild blueberries, or a crisp, sweet, and yet slightly tart apple. Watch as they bite into it and joyful astonishment spreads across their faces. You will feel good. They will feel good. With real flavor, there are no losers.

It Will Get Better

The view from the supermarket can look pretty bleak—produce aisles and meat counters filled with high-yield blandness, and calorie-rich flavor lies in aisles one through twelve. There is only one way the overall quality of food we eat will get better: if people demand it. The quality movement has revolutionized the wine and beer North Americans drink. Now let's make it happen with food.

Visit www.markschatzker.com for news on real flavor, like where to find Harry Klee's amazing tomatoes or a flavorful chicken near you.

Acknowledgments

I F YOU knock on the door of the Klee Lab long after most people have gone home for the night, it will be answered by Denise Tieman, the woman who corrals all the grad students, plants untold number of tomato seeds, and generally keeps everything moving in the right direction. Tieman's job is hardly administrative—she has cowritten numerous papers with Harry Klee and she cloned the first gene in the phenylethanol pathway, which is described on page 165. Thank you, Denise. Thank you to Hans Alborn, who worked with Joe Lewis and James Tumlinson and who identified the compounds in caterpillar spit that induce plants to synthesize and release aromatic signals to the specific natural enemy of the offending caterpillar. Thank you to Juan Villalba—not to mention all the other figures in the world of functional ecology—who was such an important part of Fred Provenza's research effort. Science is a collective effort, and in this book I have mentioned a mere handful of names. So it is with gratitude, and also reverence, that I thank the unacknowledged many.

This book may only have a single name attached to it, but it similarly represents the efforts of many. Thank you to my agent, Richard Morris, who nursed and encouraged this idea since before it was an idea, and who puts up with way too many phone calls. Thank you to Michael Szczerban for not only for commissioning and shaping *The Dorito Effect*, but for caring so much about ideas. And thank you to Millicent Bennett for proving that a book's step-editor can be every bit as nurturing as its biological editor.

What I love most about my work—more than the travel, more than the fancy meals—is the opportunity to engage with interesting thinkers and doers. Few experiences are as edifying as a conversation with the University of Illinois's renowned entomologist May Berenbaum, who walked me through the history and science of plant secondary compounds. Thanks also to John Leffingwell, for sharing so much on the early history of the flavor industry, to the University of Minnesota's Gary Reineccius, for helping me understand some of the (flagrantly ignored) minutiae of flavor labeling regulations, and to the University of Alberta's Doug Korver, a font of knowledge on the modern poultry industry. Paul Johnson, you deserve a second PhD for having to explain your PhD to me.

Every writer exists within and because of a supporting network, editors like Ted Moncreiff and Kathryn Hayward, who keep people like me employed, and friends like Pete Eshelman, Ben Auld, Allen Williams, and Jonathan Gushue. My two brothers, Adam and Erik, both seem to have been born on the same maniacal flavor quest as me. Life would be more difficult, and a good deal more boring, without you. When writing a book that covers so much science as this one does, it is an enormous advantage to have an uncle who is not only a doctor, but a biochemist as well—thank you, Ken Siren. I wouldn't be here, of course, if not for my parents. Thank you for never saying, "But wouldn't you rather be a lawyer?" and for teaching me to love real flavor. I may have hated ratatouille in grade one, but people mature—and so do palates.

This book would have never happened if not for my wonderful, beautiful wife, who, besides being the best mummy in the world as well as a very busy theater producer, is also: first reader, recipe tester, doubt deflector, therapist, life coach, cell mate and unpaid assistant. Laura, I love you. Have you seen my keys?

Bibliography

Agnew, Jeremy. *Medicine in the Old West: A History: 1850–1900.* Jefferson, N.C.: McFarland and Co., 2010.

Anthony, James C. "Epidemiology of Drug Dependence." In *Neuropsychopharmacology: The Fifth Generation of Progress,* edited by Kenneth L. Davis et al., 1556–64. Philadelphia: Lippincott Williams and Wilkins, 2002.

Atwood, S.B., Provenza, F.D., Wiedmeier, R. D., and Banner, R. E. "Influence of Free-Choice vs Mixed-Ration Diets on Food Intake and Performance of Fattening Calves." *Journal of Animal Science* 79, no. 12 (2001): 3034–40.

Ayerza, R., Coates, W., and Lauria, M. "Chia Seed (Salvia hispanica L.) as an ω-3 Fatty Acid Source for Broilers: Influence on Fatty Acid Composition, Cholesterol and Fat Content of White and Dark Meats, Growth Performance, and Sensory Characteristics." *Poultry Science* 81, no. 11 (2002): 826–37. doi: 10.2527/jas.2009-1987.

Bartoshuk, Linda M., and Klee, Harry J. "Better Fruits and Vegetables through Minireview Sensory Analysis." *Current Biology* 23, no. 9 (2013): R374–R378. doi: http://dx.doi.org/10.1016/j.cub.2013.03.038.

Bazely, Dawn R. "Carnivorous Herbivores: Mineral Nutrition and the Balanced Diet." *Trends in Ecology and Evolution* 4, no. 6 (1989): 155–6. doi: 10.1016/0169-5347(89)90115-8.

Beck, Simone, Bertholle, Louisette, and Child, Julia. *Mastering the Art of French Cooking.* New York: Knopf, 1961.

Bernays, Elizabeth A., and Singer, Michael S. "Insect Defenses: Taste Alteration and Endoparasites." *Nature* 436 (2005): 476. doi:10.1038/436476a.

Berridge, Kent, and Kringelbach, Morten. "Affective Neuroscience of Pleasure: Reward in Humans and Animals." *Psychopharmacology* 199 (2008): 457–480. DOI 10.1007/s00213-008-1099-6.

Bhatt, Arun. "Evolution of Clinical Research: A History Before and Beyond James Lind." *Perspectives in Clinical Research* 1, no. 1 (2012): 6–10.

Bittman, Mark. *How to Cook Everything.* Hoboken: Wiley Publishing, 1998.

Breslin, Paul A.S. "An Evolutionary Perspective on Food and Human Taste." *Current Biology* 23, no. 9 (2013): R409–R418. doi: 10.1016/j.cub.2013.04.010.

Bunea, Andrea, Rugina, Olivia Dumitrita, Pintea, Adela M., Sconta, Zorița, Bunea, Claudiu I., Socaciu, Carmen. "Comparative Polyphenolic Content and Antioxidant Activities of Some Wild and Cultivated Blueberries from Romania." *Notulae Botanicae Horti Agrobotanici* 39, no. 2 (2011): 17–76.

Burkhauser, Richard V., Cawley, John, and Schmeiser, Maximilian D. "The Timing of the Rise in U.S. Obesity Varies with Measure of Fatness." *Economics and Human Biology* 7, no. 3 (2009): 307–18. doi: 10.1016/j.ehb.2009.07.006.

Bushdid, C., Magnasco, M.O., Vosshall, L.B., and Keller, A. "Humans Can Discriminate More than 1 Trillion Olfactory Stimuli." *Science* 343, no. 6177 (2014): 1370-2. doi: 10.1126/science.1249168.

Catoni, C., Schaefer, H. Martin, and Peters, A. "Fruit for Health: The Effect of Flavonoids on Humoral Immune Response and Food Selection in a Frugivorous Bird." *Functional Ecology* 22, no. 4 (2008): 649-54. doi: 10.1111/j.1365-2435.2008.01400.x.

Cawley, John. "The Impact of Obesity on Wages." *The Journal of Human Resources* 39, no. 2 (2004): 451–74.

Cooper, E. A., and Funk, Casimir. "Experiments on the Causation of Beri-beri." *The Lancet* 178, no. 4601 (1911): 1266–7. doi: 10.1016/S0140-6736(01)42128-3.

Cooper, S. D. B., Kyriazakis, I., and Oldham, J. D. "The Effect of Late Pregnancy on the Diet Selections Made by Ewes." *Livestock Production Science* 40, no. 3 (1994). doi:10.1016/0301-6226(94)90094-9.

Cuellar, Steven, Karnowsky, Dan, and Acosta, Frederick. "The Sideways Effect: A Test for Changes in the Demand for Merlot and Pinot Noir Wines." *Journal of Wine Economics* 4, no.2 (2009): 219–32. doi: http://dx.doi.org/10.1017/S193143610000081X.

Cutting, Tanja M., Fisher, Jennifer O., Grimm-Thomas, Karen, and Birch, Leann L. "Like Mother, Like Daughter: Familial Patterns of Overweight Are Mediated by Mothers' Dietary Disinhibition." *The American Journal of Clinical Nutrition* 69, no. 4 (1999): 608–13.

D'Agostino, Ryan. *Eat Like a Man: The Only Cookbook A Man Will Ever Need.* San Francisco: Chronic Books, 2011.

Dal Bosco, A., Mugnai, C., Ruggeri, S., Mattioli, S., and Castellini, C. "Fatty Acid Composition of Meat and Estimated Indices of Lipid Metabolism in Different Poultry Genotypes Reared Under Organic System." *Poultry Science* 91, no. 8 (2012): 2039–45. doi: 10.3382/ps.2012-02228.

Davis, Clara M. "Results of the Self-Selection of Diets by Young Children." *The Canadian Medical Association Journal* 41, no. 3 (1939): 257–61.

Davis, Donald R. "Declining Fruit and Vegetable Nutrient Composition: What Is the Evidence?" *HortScience* 44, no. 1 (2009): 15–19.

———. "Impact of Breeding and Yield on Fruit, Vegetable, and Grain Nutrient Content." In *Breeding for Fruit Quality,* edited by Matthew A. Jenks and Penelope J. Bebeli, 127–150. Ames, Iowa: Wiley-Blackwell, 2011. doi: 10.1002/9780470959350.

———. "Trade-Offs in Agriculture and Nutrition." *Food Technology,* 59, no. 3 (2005): 120.

———, Epp, Melvin D., Riordan, Hugh D. "Changes in USDA Food Composition Data for 43 Garden Crops, 1950 to 1999." *Journal of the American College of Nutrition* 23, no. 6 (2004): 669–82.

Delgado, J., Barranco, P., and Quirce, S. "Obesity and Asthma." *Journal of Investigational Allergology and Clinical Immunology* 18, no. 6 (2008): 420–25.

Demetriades, A.K., Wallman, P.D., McGuiness, A., and Gavalas, M. C. "Low Cost, High Risk: Accidental Nutmeg Intoxication." *Emergency Medicine Journal* 22, no. 3 (2005): 223–5. doi: 10.1136/emj.2002.004168.

De Rosa, G., Moio, L., Napolitano, F., Grasso, F., Gubitosi, L., and Bordi, A. "Influence of Flavor on Goat Feeding Preferences." *Journal of Chemical Ecology* 28, no. 2 (2002): 269–81. doi: 10.1023/A:1017977906903.

Dohi, H., Yamada, A., and Fukukawa, T. "Effects of Organic Solvent Extracts from Herbage on Feeding Behavior in Goats." *Journal of Chemical Ecology* 22, no. 3 (1996): 425–30. doi: 10.1007/BF02033646.

———. "Intake Stimulants in Perennial Ryegrass (Lolium perenne L.) Fed to Sheep," *Journal of Dairy Science* 80, no. 9 (1997): 2083-6.

Dubé, Laurette et al. *Obesity Prevention: The Role of Brain and Society on Individual Behavior*. Boston: Elsevier, 2010.

Dunn, Ian C., Meddle, Simone, Wilson, Peter, Wardle, Chloe A., Law, Andy, Bishop, Valerie, Hindar, Camilla, Robertson, Graeme W., Burt, Dave W., Ellison, Stephanie J., Morrice, David M., Hocking, Paul M. "Decreased Expression of the Satiety Signal Receptor CCKAR is Responsible for Increased Growth and Body Weight During the Domestication of Chickens." *American Journal of Physiology—Endocrinology and Metabolism* 304, no. 9 (2013): E909–E921.

Dziba, L. E., Hall, J. O., and Provenza, Frederick. "Feeding Behavior of Lambs in Relation to Kinetics of 1,8 Cineole Dosed Intravenously or into the Rumen." *Journal of Chemical Ecology* 32, no. 2 (2006): 391–408.

Early, David M., and Provenza, Frederick D. "Food Flavor and Nutritional Characteristics Alter Dynamics of Food Preference in Lambs." *Journal of Animal Science* 76, no. 3 (1998): 729–34.

Esposito, Katherine, Pontillo, Alessandro, Di Palo, Carmen, Giugliano, Giovanni, Masella, Mariangela, Marfella, Raffaele, Giugliano, Dario. " Effect of Weight Loss and Lifestyle Changes on Vascular Inflammatory Markers in Obese Women." *Journal of the American Medical Association* 289, no. 14 (2003): 1799–1804. doi:10.1001/jama.289.14.1799.

Estruch, Ramón, Ros, Emilio, Salas-Salvadó, Jordi, Covas, Maria-Isabel, Corella, Dolores, Arós, Fernando, Gómez-Gracia, Enrique, Ruiz-Gutiérrez, Valentina, Fiol, Miquel, Lapetra, M.D., José, Lamuela-Raventos, Rosa Maria, Serra-Majem, Luís, Pintó, Xavier, Basora, Josep, Muñoz, Miguel Angel, Sorlí, José V., Martínez, José Alfredo, and Martínez-González, Miguel Angel. " Primary Prevention of Cardiovascular Disease with a Mediterranean Diet." *The New England Journal of Medicine*, 368 (2013): 3681279. doi: 10.1056/NEJMoa1200303.

Farmer, Fannie Merritt. *The Boston Cooking-School Cook Book*. Boston: Little, Brown, and Co., 1896.

Farmer, L.J., Perry, G.C., Lewis, P.D., Nute, G.R., Piggott, J.R., and Patterson, R. L. "Responses of Two Genotypes of Chicken to the Diets and Stocking Densities of Conventional UK and Label Rouge Production Systems-II. Sensory Attributes." *Meat Science* 47, nos. 1–2 (1997): 77–93.

FEMA 100: A Century of Great Taste. The Flavor Extract and Manufacturer's Association, 2009.

Finkelstein, Eric A., Trogdon, Justin A., Cohen, Joel W., and Dietz, William. "Annual Medical Spending Attributable to Obesity: Payer- and Service-Specific Estimates."*Health Affairs* 28, no. 5 (2009): 822–31. doi: 10.1377/hlthaff.28.5.w822.

Food Systems 25. Special Issue 01 (2010): 45–54. doi: http://dx.doi.org/10.1017/S1742170509990214.

Fryar, Cheryl D., Gu, Qiuping, and Ogden, Cynthia L. "Anthropometric reference data for children and adults: United States, 2007–2010." *National Center for Health Statistics. Vital and Health Statistics* 11, no 252 (2012): 1–40.

Furness, John B., Rivera, Leni R., Cho, Hyun-Jung, Bravo, David M., and Callaghan, Brid. "The Gut as a Sensory Organ." *Nature Reviews Gastroenterology & Hepatology* 10, no. 12 (2013): 729–40. doi:10.1038/nrgastro.2013.180.

Gawin, Frank H., and Kleber, Herbert D. "Abstinence Symptomatology and Psychiatric Diagnosis in Cocaine Abusers." *Archives of General Psychiatry* 43, no. 2 (1986): 107–13. doi:10.1001/archpsyc.1986.01800020013003.

Gearhardt, Ashley N., Yokum, Sonja, Orr, Patrick T., Stice, Eric, Corbin, William R., and Brownell, Kelly D. "Neural Correlates of Food Addiction." *Archives of General Psychiatry* 68, no. 8 (2011): 808–16. doi:10.1001/archgenpsychiatry.2011.32.

Gerrior, S., Bente, L., and Hiza, H. *Nutrient Content of the U.S. Food Supply, 1909-2000.* Home Economics Research Report No. 56. U.S. Department of Agriculture, Center for Nutrition Policy and Promotion, 2004.

Glander, Kenneth E. "The Impact of Plant Secondary Compounds on Primate Feeding Behavior." *Yearbook of Physical Anthropology* 25, no. S3 (1982): 1–18. doi: 10.1002/ajpa.1330250503.

Goodman, Gary E., Thornquist, Mark D., Balmes, John, Cullen, Mark R., Meyskens Jr., Frank L., Omenn, Gilbert S., Valanis, Barbara, and Williams Jr., James H. "The Beta-Carotene and Retinol Efficacy Trial: Incidence of Lung Cancer and Cardiovascular Disease Mortality During 6-Year Follow-up After Stopping β-Carotene and Retinol Supplements." *Journal of the National Cancer Institute* 96, no. 23 (2004): 1743–50. doi: 10.1093/jnci/djh320.

Gordy, J. Frank. "Broilers." In *American Poultry History, 1823–1973: An Anthology Overview of 150 Years*, 370–432. Edited by O. A. Hanke. Madison, Wisc.: American Printing and Publishing Inc., 1974.

Gregory, Annie R. *The Ideal Cook Book.* Chicago: American Wholesale Company, 1902.

Gugusheff, Jessica R., Ong, Zhi Yi, and Muhlhausler, Beverly S. "A Maternal 'Junk-Food' Diet Reduces Sensitivity to the Opioid Antagonist Naloxone in Offspring Postweaning." *The FASEB Journal* 27 (2013): 1275–84. doi: 10.1096/fj.12-217653.

Hagen, Edward H., Roulette, Casey J., and Sullivan, Roger J. "Explaining Human Recreational Use of 'Pesticides': The Neurotoxin Regulation Model of Substance Use vs. the Hijack Model and Implications for Age and Sex Differences in Drug Consumption." *Frontiers in Psychiatry* 4 (2014): 114. doi: 10.3389/fpsyt.2013.00142.

Hall, Richard L., and Oser, Bernard L. "Recent Progress in the Consideration of Flavoring Ingredients Under the Food Additives Ammendment: *GRAS Substances*." *Food Technology* 19, no. 2 (1965): 151–97.

Halliwell, Barry. Polyphenols: Antioxidant Treats for Healthy Living or Covert

Toxins?" *Jounral of the Science of Food and Agriculture* 86 (2006): 1992–5. doi: 10.1002/jsfa.2612.

Hardy, K., Buckley, S., Collins, M. J., Estalrrich, A., Brothwell, D., Copeland, L., García-Tabernero, A., García-Vargas, S., de la Rasilla, M., Lalueza-Fox, C., Huguet, R., Bastir M., Santamaría, D., Madella, M., Wilson, J., Cortés, A. F., Rosas, A. "Neanderthal Medics? Evidence for Food, Cooking, and Medicinal Plants Entrapped in Dental Calculus." *Naturwissenschaften* 99, no. 8 (2012): 617–26. doi: 10.1007/s00114-012-0942-0.

Hartmann, Thomas. "From Waste Products to Ecochemicals: Fifty Years Research of Secondary Metabolism." *Phytochemistry* 68, no. 22–24 (2007): 2831–46.

Hazan, Marcella. *Marcella Says*. New York: Harper Collins, 2004.

He, Weiping, and Huang, Baokang. "A Review of Chemist and Bioactivities of a Medicinal Spice: *Foeniculum Vulgare*." *Journal of Medicinal Plants Research* 5, no. 16 (2011): 3595–3600.

Hellmich, Nancy. "Weight Watchers Founder Jean Nidetch Shares Her Start." *USA Today,* March 22, 2010.

Henderson, J.T. *A Manual on Poultry*. Atlanta: J. P. Harrison and Co., 1883.

Huffman, Michael, Gotoh, Shunji, Izutsu, Daisuke, Koshimizu, Koichi, and Kalunde, Mohamedi Seifu. "Further Observations on the Use of the Medicinal Plant, *Vernonia Amygdalina* (Del), by a Wild Chimpanzee, Its Possible Effect on Parasite Load, and Its Phytochemistry." *African Study Monographs* 14, no. 4 (1993): 227–40.

Huffman, Michael A. "Self-Medicative Behaviour in the African Great Apes: An Evolutionary Perspective into the Origins of Human Traditional Medicine." *BioScience* 51, no. 8 (2001): 651–61. doi: http://dx.doi.org/10.1641/0006 -3568(2001)051[0651:SMBITA]2.0.CO;2.

Jansky, Shelley H. "Potato Flavor." In *Handbook of Fruit and Vegetable Flavors*, edited by Y. H. Hui, 935–946. Hoboken: John Wiley & Sons, 2010.

Janssen, Sara, Laermans, Jorien, Verhuilst, Pieter-Jan, Thijs, Theo, Tack, Jan, and Depoortere, Inge. "Bitter Taste Receptors and α-gustducin Regulate the Secretion of Ghrelin with Functional Effects on Food intake and Gastric Emptying." *Proceedings of the National Academy of Sciences* 108, no. 5 (2011): 2094–9. doi: 10.1073/pnas.1011508108.

Jia, H., and Lubetkin, Erica. "Trends in Quality-Adjusted Life-Years Lost Contributed by Smoking and Obesity." *American Journal of Preventive Medicine* 38, no. 2 (2010): 138–44. doi: 10.1016/j.amepre.2009.09.043.

Johns, Timothy. *The Origins of Human Diet and Medicine*. Tucson: The University of Arizona Press, 1990.

Johnson, Paul M., and Kenny, Paul. "Dopamine D2 Receptors in Addiction-like Reward Dysfunction and Compulsive Eating in Obese Rats." *Nature Neuroscience* 13 (2010): 635–41. doi: 10.1038/nn.2519.

Judge, Timothy A., and Cable, Daniel M. "When It Comes to Pay, Do the Thin Win? The Effect of Weight on Pay for Men and Women." *Journal of Applied Psychology* 96, no. 1 (2011): 95–112. doi: 10.1037/a0020860.

Juhnke, J., Mill, J. O., Provenza, F. D., and Villalba, J. J. "Preference for Condensed Tannins by Sheep in Response to Challenge Infection with *Haemonchus contortus*." *Veterinary Parasitology* 188 (2012): 104–14. doi: 10.1016/j.vetpar.2012.02.015.

Karsten, H.D., Patterson, P.H., Stout, R., and Crews, G. "Vitamins A, E and Fatty Acid Composition of the Eggs of Caged Hens and Pastured Hens." *Renewable Agriculture and Food Systems* 25, no. 1 (2010): 45–54. http://dx.doi.org/10.1017/S1742170509990214.

Kaspar, Kerrie L., Park, Jean Soon, Brown, Charles R., Mathison, Bridget D., Navarre, Duroy A., and Chew, Boon P. "Pigmented Potato Consumption Alters Oxidative Stress and Inflammatory Damage in Men." *The Journal of Nutrition and Disease* 141, no.1 (2011): 108–11. doi: 10.3945/jn.110.128074.

Keller, Thomas. *Ad Hoc at Home.* New York: Artisan Books, 2009.

Kenny, Paul J. "Common Cellular and Molecular Mechanisms in Obesity and Drug Addiction." *Nature Reviews Neuroscience* 12 (2011): 638–51. doi:10.1038/nrn3105.

———. "Reward Mechanisms in Obesity: New Insights and Future Directions." *Neuron* 69, no. 4 (2011): 664–79. doi: 10.1016/j.neuron.2011.02.016.

Kessler, David A. *The End of Overeating: Taking Control of the Insatiable North American Appetite.* Toronto: McClelland and Stewart, 2009.

Kim, DaeHwan, and Leigh, John Paul. "Estimating the Effects of Wages on Obesity." *Journal of Occupational and Environmental Medicine* 52, no. 5 (2010): 495–500. doi: 10.1097/JOM.0b013e3181dbc867.

Klee, Harry J., and Tieman, Denise M. "Genetic Challenges of Flavor Improvement in Tomato." *Trends in Genetics* 29, no. 4 (2013): 257–62. doi: 10.1016/j.tig.2012.12.003.

Koloverou, Efi, Esposito, Katherine, Giugliano, Dario, and Panagiotakos, Demosthenes. "The Effect of Mediterranean Diet on the Development of Type 2 Diabetes Mellitus: A Meta-Analysis of 10 Prospective Studies and 136,846 Participants." *Metabolism* 63, no. 7 (2014): 90311. doi: 10.1016/j.metabol.2014.04.010.

Kopelman, P. "Health Risks Associated with Overweight and Obesity." *Obesity Reviews* 8, s1 (2007): 13–17. doi: 10.1111/j.1467-789X.2007.00311.x.

Kucharik, Christopher J., and Ramankutty, Navin. "Trends and Variability in U.S. Corn Yields Over the Twentieth Century." *Earth Interactions* 9, no. 1 (2005): 1–29.

Kyriazakis, I., and Oldham, J. D. "Diet Selection in Sheep: The Ability of Growing Lambs to Select a Diet That Meets Their Crude Protein (Nitrogen × 6.25) Requirements." *British Journal of Nutrition* 69, no. 3 (May, 1993): 617–29.

Lee, C.Y., Lee, B.D. Na, J-C. and An, G. "Carotenoid Accumulation and Their Antioxidant Activity in Spent Laying Hens as Affected by Polarity and Feeding Period." *Asian-Australian Journal of Animal Sciences* 23, no. 6 (2010): 799-805. doi: http://dx.doi.org/10.5713/ajas.2010.90296.

Leffingwell, John C., Young, Harvey J., and Bernasek, Edward. *Tobacco Flavoring for Smoking Products.* Winston-Salem, N.C.: R. J. Reynolds Tobacco Company, 1972.

Lind, James. *A Treatise on the Scurvy: In Three Parts.* London: A. Millar, 1753.

Lucantonio, Federica, Stainaker, Thomas A., Shaham, Yavin, Niv, Yael, and Schoenbaum, Geoffrey. "The Impact of Orbitofrontal Dysfunction on Cocaine Addiction." *Nature Neuroscience* 15 (2012): 358–66. doi:10.1038/nn.3014.

Ludy, Mary-Jon, and Mattes, Richard D. "The Effects of Hedonically Acceptable Red Pepper Doses on Thermogenesis and Appetite." *Physiology and Behavior* 102, nos. 3–4 (2011): 251–8. doi: 10.1016/j.physbeh.2010.11.018.

Luppino, Floriana S., de Wit, Leonore M., Bouvy, Paul F., Stijnen, Theo, Cuijpers, Pim, Penninx, Brenda W. J. H., Zitman, Frans G. "Overweight, Obesity, and Depression: A Systematic Review and Meta-analysis of Longitudinal Studies." *Archives of General Psychiatry* 67, no. 3 (2010): 220–9. doi: 10.1001/archgenpsychiatry.2010.2.

Manthey, John A., and Perkins-Veazie, Penelope. "Levels of β-Carotene, Ascorbic Acid, and Total Phenols in the Pulp of Five Commercial Varieties of Mango (Mangifera indica L.)." *Proceedings of the Florida State Horticultural Society* 122 (2009): 303–7.

Marcy, Randolph B. *The Prairie Traveler: A Handbook for Overland Expeditions.* London: Trubner & Company, 1863.

Mayer, Anne-Marie. "Historical Changes in the Mineral Content of Fruits and Vegetables." *British Food Journal* 99, no. 6 (1997): 207–11.

Menz, Kenneth M., and Fleming, Euan M. "Economic Prospects for Vanilla in the South Pacific." *ACIAR Technical Reports Series* 11 (1989): 1–14.

Molan, A., Lila, M.A., and Mawson, J. "Satiety in Rats Following Blueberry Extract Consumption Induced by Appetite-Suppressing Mechanisms Unrelated to in Vitro or in Vivo Antioxidant Capacity." *Food Chemistry* 107, no. 3 (2008): 103944. doi: 10.1016/j.foodchem.2007.09.018.

Moraes, C.M. de, Lewis, W.J., Paré, P.W., Alborn, H.T., and Tumlinson, James. "Herbivore-infested Plants Selectively Attract Parasitoids." *Nature* 393 (June 11, 1898): 570–3. doi: 10.1038/31219.

Moran, Andrew W., Al-Rammahi, Miran, Arora, Daleep K., Batchelor, Daniel J., Coulter, Erin A., Daly, Kristian, Ionescu, Catherine, Bravo, David, and Shirazi-Beechey, Soraya P. "Expression of Na⁺/glucose Co-Transporter 1 (SGLT1) in the Intestine of Piglets Weaned to Different Concentrations of Dietary Carbohydrate." *British Journal of Nutrition* 104, no. 5 (2010): 647–55. doi: 10.1017/S0007114510000954.

Morris, Cindy E., and Sands, David C. "The Breeder's Dilemma—Yield or Nutrition?" *Nature Biotechnology* 24 (2006): 1078–80.

Morris, Wayne, Shepherd, Tom, Verrall, Susan R., McNicol, James W., and Taylor, Mark A. "Relationships Between Volatile and Non-Volatile Metabolites and Attributes of Processed Potato Flavour." *Phytochemistry* 71, nos. 14–15 (2010): 1765–73. doi: 10.1016/j.phytochem.2010.07.003.

Moss, Michael. *Salt Sugar Fat: How the Food Giants Hooked Us.* New York: Random House, 2013.

Mote, Travis E., Villalba, Juan J., and Provenza, Frederick H. "Sequence of Food Presentation Influences Intake of Foods Containing Tannins and Terpenes." *Applied Animal Behavior Science* 113, nos. 1–3 (2008): 57–68. doi: 10.1016/j.applanim.2007.10.003.

Murphy, C., Cain, W.S., and Bartoshuk, Linda M. "Mutual Action of Taste and Olfaction." *Sense Processes* 1, no. 3 (1977): 204–11.

Myers, A., and Rosen. J.C. "Obesity Stigmatization and Coping: Relation to Mental Health Symptoms, Body Image, and Self-Esteem." *International Journal of Obesity* 23, no. 3 (1999): 221–30.

Naim, Michael, Brand, Joseph G., Kare, Morley R., and Carpenter, Richard G. "Intake, Weight Gain and Fat Deposition in Rats Fed Flavored, Nutritionally Controlled Diets in a Multichoice ('Cafeteria') Design." *The Journal of Nutrition* 115, no. 11 (1985): 1447–58.

Nakamura, Yuko, Goto, Takuko K., Tokumori, Kenzo, Yoshiura, Takashi, Kobayashi, Koji, Nakamura, Yasuhiko, Honda, Hiroshi, Ninomiya, Yuzo, and Yoshiura, Kazunori. "Localization of Brain Activation by Umami Taste in Humans." *Brain Research* 1406 (2011): 18–29. doi: 10.1016/j.brainres.2011.06.029.

Ng, Tze Pin, Chiam, Peak-Chiang, Lee, Theresa, Chua, Hong-Choon, Lim, Leslie, and Ee-Heok, Kua. "Curry Consumption and Cognitive Function in the Elderly." *American Journal of Epidemiology* 164, no. 9 (2006): 898–906.

Nidetch, Jean, and Heilman, Joan Rattner. *The Story of Weight Watchers.* New York: W/W Twentyfirst Corp.; [Cleveland]: Distributed by the New American Library in association with the World Pub. Co., 1970.

Nollet, Leo M. L. *Handbook of Meat, Seafood and Poultry Quality.* Ames, Iowa: Wiley-Blackwell, 2012.

Norton, Jeanette Young. *Mrs. Norton's Cook-Book: Selecting, Cooking and Serving for the Home Table.* New York: G.P. Putnam's Sons, 1917.

Ogden, Cynthia L., Fryar, Cheryl D., Carroll, Margaret D., and Flegal, Katherine M. "Mean Body Weight, Height, and Body Mass Index, United States 1960–2002." *Advance Data from Vital and Health Statistics* 347 (2004): 1–17.

Ogden, Cynthia L., Carroll, Margaret D., Kit, Brian K., and Flegal, Katherine M. "Prevalence of Obesity in the United States, 2009–2010." *NCHS Data Brief* 82 (2012).

———. "Prevalence of Childhood and Adult Obesity in the United States 2011–2012." *The Journal of the American Medical Association* 311, no. 8 (2014): 806–14. doi:10.1001/jama.2014.732.

Ohsu, Takeaki, Amino, Yusuke, Nagasaki, Hiroaki, Yamanaka, Tomohiko, Takeshita, Sen, Hatanaka, Toshihiro, Maruyama, Yutaka, Miyamura, Naohiro, and Eto, Yuzuru. "Involvement of the Calcium-sensing Receptor in Human Taste Perception." *The Journal of Biological Chemistry* 285, no. 2 (2010): 1016–22. doi: 10.1074/jbc.M109.029165.

Pancosma, Research News No. 16: "Sucram," March 2011.

Plassmann, Hilke, O'Doherty, John, Shiv, Baba, and Rangel, Antonio. "Marketing Actions Can Modulate Neural Representations of Experienced Pleasantness." *Proceedings of the National Academy of Sciences* 105, no. 3 (2008): 1050–4. doi: 10.1073/pnas.0706929105.

Ponte, P. I., Prates, J. A., Crespo, J. P., Crespo, D. G., Mourão, J. L., Alves, S. P., Bessa, R. J., Chaveiro-Soares, M. A., Ferreira, L. M., and Fontes, C. M. "Improving the Lipid Nutritive Value of Poultry Meat Through the Incorporation of a Dehydrated Leguminous-Based Forage in the Diet for Broiler Chicks." *Poultry Science* 87, no. 8 (2008): *doi: 10.3382/ps.2007-004461587–94.*

———. "Restricting the Intake of a Cereal-Based Feed in Free-Range-Pastured Poultry: Effects on Performance and Meat Quality." *Poultry Science* 87, no. 10 (2008): 2032–42. doi: 10.3382/ps.2007-00522.

Powell, Ann L.T., Nguyen, Cuong V., Hill, Theresa, Cheng, KaLai Lam, Figueroa-Balderas, Rosa, Aktas, Hakan, Ashrafi, Hamid, Pons, Clara, Fernández-Muñoz, Rafael, Vicente, Ariel, Lopez-Baltazar, Javier, Barry, Cornelius S., Liu, Yongsheng,

Chetelat, Roger, Granell, Antonio, Van Deynze, Allen, Giovannoni, James J., and Bennett, Alan B. "Uniform Ripening Encodes a Golden 2-like Transcription Factor Regulating Tomato Fruit Chloroplast Development." *Science* 336, no. 6089 (2012): 1711–15. doi: 10.1126/science.1222218.

Prescott, John. "Chemosensory Learning and Flavour: Perception, Preference and Intake." *Physiology and Behaviour* 107, no. 4 (2012): 553-9. doi: 10.1016/j.physbeh. 2012.04.008.

Provenza, Frederick D., and Malecheck, J. C. "Diet Selection by Domestic Goats in Relation to Blackbrush Twig Chemistry." *Journal of Applied Ecology* 21, no. 3 (1984): 831–41.

Provenza, Frederick D., Burritt, Elizabeth A., Clausen, T. P., Bryant, J.P., Reichardt, P.B., and Distel, Roberto A. "Conditioned Flavor Aversion: A Mechanism for Goats to Avoid Condensed Tannins in Blackbrush." *The American Naturalist* 136, no. 6 (1990): 810–28.

Psaltopoulou, Theodora, Sergentanis, Theodoros N., Panagiotakos, Demosthenes B., Sergentanis, Ioannis N., Kosti, Rena, and Scarmeas, Nikolaos. "Mediterranean Diet, Stroke, Cognitive Impairment, and Depression: A Meta-Analysis." *Annals of Neurology* 74, no. 4 (2013): 580–91.

Puglisi, Michael J., Mutungi, Gisella, Brun, Pierre J., McGrane, Mary M., Labonte, Cherise, Volek, Jeff S., and Fernandez, Maria Luz. "Raisins and Walking Alter Appetite Hormones and Plasma Lipids by Modifications in Lipoprotein Metabolism and Up-Regulation of the Low-Density Lipoprotein Receptor." *Metabolism* 58, no.1 (2009): 120–8. doi: 10.1016/j.metabol.2008.08.014.

Raghavendra, R.H., and Akhilender Naidu, K. "Spice Active Principles as the Inhibitors of Human Platelet Aggregation and Thromboxane Biosynthesis." *Prostaglandins, Leukotrienes and Essential Fatty Acids* 81, no. 1 (2009): 73–8. doi: 10.1016/j. plefa.2009.04.009.

Rahayu, H.S., Iman, Zulkifli, I., Vidyadaran, M.K., Alimon, A.R., and S. A. Babjee, S. A. "Carcass Variables and Chemical Composition of Commercial Broiler Chickens and the Red Jungle Fowl." *Asian-Australian Journal of Animal Sciences* 21, no. 9 (2008): 1376–82. doi: http://dx.doi.org/10.5713/ajas.2008.70662.

Ramaswamy, H. S., and Richards, J. F. "Flavor of Poultry Meat—A Review." *Canadian Institute of Food Science and Technology Journal* 15, No. 11 (1982): 7–18.

Reineccius, Gary. *Flavor Chemistry and Technology*. Boca Raton: Taylor & Francis, 2006.

Report of the Committee appointed by the Lords Commissioners of the Admiralty, to enquire into the causes of the outbreak of scurvy in the recent Arctic expedition: the adequacy of the provision made by the Admiralty in the way of food, medicine, and medical comforts: and the propriety of the orders given by the commander of the expedition for provisioning the sledge parties. London: H.M. Stationery Office, 1877.

Robins, Lee N. "Vietnam Veterans' Rapid Recovery from Heroin Addiction: A Fluke or Normal Expectation?" *Addiction* 88 (1993): 104154.

———, Davis, Darlene H., and Goodwin, Donald W. "Drug Use by U.S. Army Enlisted Men in Vietnam: A Follow-Up On Their Return Home." *American Journal of Epidemiology* 99, no. 4 (1974): 235–49.

Rolls, Barbara J. "Sensory-Specific Satiety." *Nutrition Reviews* 44, no. 3 (1986): 93–101. doi 10.1111/j.1753-4887.1986.tb07593.x.

Rolls, Edmund T. "The Representation of Umami Taste in the Taste Cortex." *The Journal of Nutrition* 130, 4S (Suppl) (2000): 960S–5S.

———. "Sensory Processing in the Brain Related to the Control of Food Intake." *Proceedings of the Nutrition Society* 66, no. 1 (2007): 96–112.

———. "Taste, Olfactory, and Food Texture Processing in the Brain, and the Control of Food Intake." *Physiology and Behavior* 85, no. 1 (2005): 45–56.

Roman Farm Management; The Treatises of Cato and Varro Done into English, with Notes of Modern Instances by a Virginia Farmer. Translated and edited by Fairfax Harrison. New York: MacMillan, 1918.

Rombauer, Irma S., Becker, Marion Rombauer, and Becker, Ethan. *Joy of Cooking.* New York: Simon and Schuster, 1997.

Rorer, S.T. *Good Cooking.* Philadelphia: The Curtis Publishing Company, 1898.

Rubin, Carl-Johan, Zody, Michael C., Eriksson, Jonas, Meadows, Jennifer R. S., Sherwood, Ellen, Webster, Matthew T., Jiang, Lin, Ingman, Max, Sharpe, Ted, Ka, Sojeong, Hallböök, Finn, Besnier, Francois, Carlborg, Örjan, Bed'hom, Bertrand, Tixier-Boichard, Michèle, Jensen, Per, Siegel, Paul, Lindblad-Toh, Kerstin, and Andersson, Leif. "Whole-genome Resequencing Reveals Loci under Selection during Chicken Domestication." *Nature* 464 (March 25, 2010): 587–91. doi:10.1038/nature08832.

Rudenga, K.J., and Small, D.M. "Amygdala Response to Sucrose Consumption Is Inversely Related to Artificial Sweetener Use." *Appetite* 58, no.2 (2012): 504–7. doi: 10.1016/j.appet.2011.

Saul, Hayley, Madella, Marco, Fischer, Anders, Glykou, Aikaterini, Hartz, Sönke, and Craig, Oliver E. "Phytoliths in Pottery Reveal the Use of Spice in European Prehistoric Cuisine." *PLOS ONE* 8, no. 8 (2013): 1–5. doi: 10.1371/journal.pone.0070583.

Sclafani, Anthony, and Ackroff, Karen. "Role of Gut Nutrient Sensing in Stimulating Appetite and Conditioning Food Preference." *American Journal of Physiology—Regulatory, Integrative, and Comparative Physiology* 302, no. 10 (2012): R1119–33. doi: 10.1152/ajpregu.00038.2012.

Seefeldt, Steven S. "Consequences of Selecting Rambouillet Ewes for Mountain Big Sagebrush (Artemisia tridentata ssp. vaseyana) Dietary Preference." *Rangeland Ecology Management* 58, no. 4 (2005): 380–4.

Shepherd, Gordon M. *Neurogastronomy: How the Brain Creates Flavor and Why It Matters.* New York: Columbia University Press, 2012.

Simone-Finstrom, Michael D., and Spivak, Maria. "Increased Resin Collection after Parasite Challenge: A Case of Self-Medication in Honey Bees?" *PLOS ONE* 7, no. 3 (2012): 1. doi: 10.1371/journal.pone.0034601.

Singh, Sukhvinder Pal, and Singh, Zora. "Postharvest Cold Storage-Induced Oxidative Stress in Japanese Plums (Prunus Salicina Lindl. cv. Amber Jewel) in Relation to Harvest Maturity." *Australian Journal of Crop Science* 7, no. 3 (2013): 391.

Singletary, Keith. "Cinnamon: Overview of Health Benefits." *Nutrition Today* 43, no. 6 (2008): 263–6. doi: 10.1097/01.NT.0000342702.19798.fe.

———. "Turmeric: An Overview of Potential Health Benefits." *Nutrition Today* 45, no. 5 (2010): 216–25. doi: 10.1097/NT.0b013e3181f1d72c.

Sinha, Arun K., Sharma,Upendra K., and Sharma, Nandini. "A Comprehensive Review on Vanilla Flavor: Extraction, Isolation and Quantification of Vanillin and Others Constituents." *International Journal of Food Sciences and Nutrition* 59, no. 4 (2008): 299–326. doi: 10.1080/09687630701539350.

Snitker, Soren, Fujishima, Yoshiyuki, Shen, Haiqing, Ott, Sandy, Pi-Sunyer, Xavier, Furuhata, Yasufumi, Sato, Hitoshi, and Takahashi, Michio. "Effects of Novel Capsinoid Treatment on Fatness and Energy Metabolism in Humans: Possible Pharmacogenetic Implications." *The American Journal of Clinical Nutrition* 89, no.1 (2009): 45–50. doi: 10.3945/ajcn.2008.26561.

Sohm, B., Cenizo, V., André, V., Zahouani, H., Pailler-Mattei, C., and Vogelgesang, B. "Evaluation of the Efficacy of a Dill Extract in Vitro and in Vivo." in *International Journal of Cosmetic Science* 33, no. 2 (2011): 157–63. doi: 10.1111/j.1468-2494.2010.00606.

Speth, John D., and Spielmann, Katherine A. "Energy Source, Protein Metabolism, and Hunter-Gatherer Subsistence Strategies." *Journal of Anthropological Archaeology* 2 (1983): 1–31. http://dx.doi.org/10.1016/0278-4165(83)90006-5.

Spring, Bonnie, Schneider, Kristin, Smith, Malaina, Kendzor, Darla, Appelhans, Bradley, Hedeker, Donald, and Pagoto, Sherry. "Abuse Potential of Carbohydrates for Overweight Carbohydrate Cravers." *Psychopharmacology* 197, no. 4 (2008): 637–47. doi: 10.1007/s00213-008-1085-z.

Stefansson, Vilhjalmur. *The Fat of the Land: Enlarged Edition of Not By Bread Alone.* New York: Macmillan, 1956.

Stein, Stephanie, Lamos, Elizabeth, Quartuccio, Michael, Chandraskaran, Sruti, Ionica, Nicole, and Steinle, Nanette. "Food Intake and Food Preference." In *Diet Quality: An Evidence-Based Approach.* Vol. 1. Edited by Victor R. Preedy, Lan-Anh Hunter, Vinood B. Patel, 13–25. New York: Springer Science+Business Media, 2013.

Stice, Eric, Spoor, Sonja, Bohon, Cara, Veldhuizen, Marga G., and Small, Dana M. "Relation of Reward from Food Intake and Anticipated Food Intake to Obesity: A Functional Magnetic Resonance Imaging Study." *Journal of Abnormal Psychology* 117, no. 4 (2007): 924–35. doi: 10.1037/a0013600.

Strauss, Richard S. "Childhood Obesity and Self-Esteem." *Pediatrics* 105, no. 1 (2000): e15.

Sturm, Roland, and Wells, Kenneth B. "Does Obesity Contribute as Much to Morbidity as Poverty or Smoking?" *Public Health* 115, no. 3 (2001): 225–35.

Sun, T., Long, R. J., Liu, Z. Y., Ding, W. R., and Zhang, Y. "Aspects of Lipid Oxidation of Meat from Free-Range Broilers Consuming a Diet Containing Grasshoppers on Alpine Steppe of the Tibetan Plateau." *Poultry Science* 91, no. 1 (2012): 224–31. doi: 10.3382/ps.2011-01598.

Swithers, Susan E. "Artificial Sweeteners Produce the Counterintuitive Effect of Inducing Metabolic Derangements." *Trends in Endocrinology and Metabolism* 24, no. 9 (2013): 431–41.

Thorn, Charles Embree. *The Complete Poultry Book.* Springfield: Mast, Crowell and Kirkpatrick, 1882.

Tieman, Denise, Bliss, Peter, McIntyre, Lauren M., Blandon-Ubeda, Adilia, Bies, Dawn, Odabasi, Asli Z., Rodriguez, Gustavo R., van der Knaap, Esther, Taylor,

Mark G., Goulet, Charles, Mageroy, Melissa H., Snyder, Derek J., Colquhoun, Thomas, Moskowitz, Howard, Clark, David G., Sims, Charles, Bartoshuk, Linda, and Klee, Harry J. "The Chemical Interactions Underlying Tomato Flavor Preferences." *Current Biology* 22, no. 11 (2012): 1035-39. doi: http://dx.doi.org/10.1016/j.cub.2012.04.016.

Treit, Dallas, Spetch, Marcia L., and Deutsch, J. A. "Variety in the Flavor of Food Enhances Eating in the Rat: A Controlled Demonstration." *Physiology and Behavior* 30, no. 2 (1983): 207–11.

Trotter, Thomas. *Observations on the Scurvy.* London: T. Longman and J. Watts, 1792.

Turlings, Ted C., and Tumlinson, James H. "Systemic Release of Chemical Signals by Herbivore-Injured Corn." *Proceedings of the National Academy of Sciences* 89, no. 17 (1992): 8399–402.

———, and Lewis, W. Joe. "Exploitation of Herbivore-Induced Plant Odors by Host-Seeking Parasitic Wasps." Science 250, no. 4985 (1990): 1251–3. doi: *10.1126/science.250.4985.1251.*

Turlings, Ted C., Loughrin, John H., McCall, Philip J., Röse, Ursula S. R., Lewis, W. Joe, and Tumlinson, James H. "How Caterpillar-Damaged Plants Protect Themselves by Attracting Parasitic Wasps." *Proceedings of the National Academy of Sciences* 92 (1995): 4169–74.

Ueda, Y., Yonemitsu, M., Tsubuku, T., Sakaguchi, M., and Miyajima, R. "Flavor Characteristics of Glutathione in Raw and Cooked Foodstuffs." *Bioscience, Biotechnology, and Biochemistry* 61, no. 12 (1997): 1977–80.

"The Ups and Downs of Folic Acid Fortification." *Harvard Women's Health Watch* (March 2008). http://www.health.harvard.edu/newsweek/the-ups-and-downs-of-folic-acid-fortification.htm.

Villalba, Juan J., and Provenza, Frederick D. "Nutrient-Specific Preferences by Lambs Conditioned with Intraruminal Infusions of Starch, Casein, and Water." *Journal of Animal Science* 77, no.2 (1999): 378–87.

———. "Postingestive Feedback from Starch Influences the Ingestive Behaviour of Sheep Consuming Wheat Straw." *Applied Animal Behaviour Science* 66, nos. 1–2 (2000): 49–63. doi: 10.1016/S0168-1591(99)00081-7.

———, and Hall, J. O. "Learned Appetites for Calcium, Phosphorus, and Sodium in Sheep." *Journal of Animal Science* 86, no. 3 (2008): 738–47.

———, Hall, J. O., and Peterson, C. "Phosphorus Appetite in Sheep: Dissociating Taste from Postingestive Effects." *Journal of Animal Science* 84, no. 8 (2005): 2213–23. doi: 10.2527/jas.2005-634.

———, and Shaw, Ryan. "Sheep Self-Medicate When Challenged with Illness-nducing Foods." *Animal Behaviour* 71, no. 5 (2005): 1131-9. doi: 10.1016/j.anbehav.2005.09.012.

Vinson, Joe A., Demkosky, Cheryl A., Navarre, Duroy A., and Smyda, Melissa A. "High-Antioxidant Potatoes: Acute in Vivo Antioxidant Source and Hypotensive Agent in Humans after Supplementation to Hypertensive Subjects." *Journal of Agriculture and Food Chemistry* 60, no.27 (2012): 6749–54. doi: 10.1021/jf2045262.

Walter, Richard, *A Voyage round the World 1740–1744,* by George Anson, compiled from his papers by Richard Walter, Chaplain to the *Centurion.* London: Society for Promoting Christian Knowledge, 1838.

Wang, Yiqun, Lehane, Catherine, Ghebremeskel, Kebreab, and Crawford, Michael A. "Modern Organic and Broiler Chickens Sold for Human Consumption Provide More Energy from Fat Than Protein." *Public Health Nutrition* 13, no. 3 (2010): 400–8. doi: 10.1017/S1368980009991157.

Wideman, R. F., Rhoads, D. D., Erf, G. F., and Anthony, N. B. "Pulmonary Arterial Hypertension (ascites syndrome) in Broilers: A Review." *Poultry Science* 92, no. 1 (2013): 64–83. doi: 10.3382/ps.2012-02745.

Wiedmeiera, R.W., Villalba, J. J., Summersa, A., and Provenza, Frederick D. "Eating a High Fiber Diet During Pregnancy Increases Intake and Digestibility of a High Fiber Diet by Offspring in Cattle." *Animal Feed Science and Technology* 177, nos. 3–4 (2012): 144–51. doi: http://dx.doi.org/10.1016/j.anifeedsci.2012.08.006.

Wilson, Emily A., and Demmig-Adams, Barbara. "Antioxidant, Anti-inflammatory, and Antimicrobial Properties of Garlic and Onions." *Nutrition and Food Science* 37, no. 3 (2007): 178–83. doi: 10.1108/00346650710749071.

Wu, Trang-Tiau, Tsai, Chia-Wen, Yao, Hsieng-Tsung, Lii, Chong-Kuei, Chen, Haw-Wen, Wu, Yu-Ling, Chen, Pei-Yin, and Liu, Kai-Li. "Suppressive Effects of Extracts from the Aerial Part of Coriandrum Sativum L. on LPS-Induced Inflammatory Responses in Murine RAW 264.7 Macrophages." *Journal of the Science of Food and Agriculture* 90, no. 11 (2010): 1846–54. doi: 10.1002/jsfa.4023.

Yeomans, Martin R. "Flavour-Nutrient Learning in Humans: An Elusive Phenomenon?" *Physiology and Behaviour* 106, no. 3 (2012): 345–55. doi: 10.1016/j.physbeh.2012.03.013.

Yoshioka, Mayumi, St-Pierre, Sylvie, Drapeau, Vicky, Dionne, Isabelle, Doucet, Eric, Suzuki, Masashige, and Tremblay, Angelo. "Effects of Red Pepper on Appetite and Energy Intake." *British Journal of Nutrition* 82, no. 2 (1999): 115–23.

Ziauddeen, Hisham, Farooqi, I. Sadaf, and Fletcher, Paul C. "Obesity and the Brain: How Convincing Is the Addiction Model?" *Nature Reviews Neuroscience* 13 (2012): 279–86.

Notes

One: "Things" and "Flavors"

3 **In the early autumn of 1961:** Almost all the details of Jean Nidetch's story are taken from Nidetch's own *The Story of Weight Watchers* (New York: Twentyfirst Corp., 1970).

5 **Within five years, 297 classes:** Nan Ickeringill, "Weight Watchers: Talking Their Way Out of Obesity," *New York Times*, March 20, 1967.

5 **bought her business for $72 million:** Obituary for Albert Lippert, a Weight Watchers cofounder, *New York Times*, March 3, 1998.

6 **According to the Centers for Disease Control:** All obesity statistics are taken from "Prevalence of Overweight, Obesity, and Extreme Obesity Among Adults: United States, Trends 1960–1962 Through 2009–2010" (Atlanta: CDC National Center for Health Statistics, 2013), as well as Cynthia L. Ogden et al., "Prevalence of Childhood and Adult Obesity in the United States, 2011–2012," *Journal of the American Medical Association* 311, no. 8 (2014): 806–14. Prior to 1988, the NHES and NHANES data covers U.S. adults age twenty to seventy-four. The 2012 data I cite covers adults twenty and older. I am grateful to the CDC's Cynthia Ogden for her assistance in interpreting these figures.

6 **obesity is measured by skin fold:** R. V. Burkhauser, J. Cawley, and M. D. Schmeiser, "Timing of the Rise in U.S. Obesity Varies with Measure of Fatness," *Economics and Human Biology* 7, no. 3 (2009): 307–18.

6 **sold-out Pirates-Yankees World Series game:** The figure is calculated for the October 8 game, which the Yankees won 10–0 and whose attendance, according to retrosheet.org, was 70,001. The figure of 4,500 assumes an equivalent attendance.

7 **Ninety million Americans:** According to the CDC's "Prevalence of Obesity in the United States, 2009–2010."

7 **increased risk of asthma:** This is not an exhaustive list. P. Kopelman, "Health Risks Associated with Overweight and Obesity," *Obesity Review* 8, Suppl. 1 (2007), covers many of the associated risks. See also J. Delgado, P. Barrranco, and S. Quirce, "Obesity and Asthma," *Journal of Investiga-*

tional Allergology and Clinical Immunology 18, no. 6 (2008); John Cawley, "The Impact of Obesity on Wages," *Journal of Human Resources* 39, no. 2 (2004); D. Kim and J. P. Leigh, "Estimating the Effects of Wages on Obesity," *Journal of Occupational and Environmental Medicine* 52, no. 5 (2010); T. A. Judge and D. M. Cable, "When It Comes to Pay, Do the Thin Win? The Effect of Weight on Pay for Men and Women," *Journal of Applied Psychology* 96, no. 1 (2011); E. A. Finkelstein et al., "Annual Medical Spending Attributable to Obesity: Payer- and Service-Specific Estimates," *Health Affairs* 28, no. 5 (2009); F. S. Luppino et al., "Overweight, Obesity, and Depression: A Systematic Review and Meta-Analysis of Longitudinal Studies," *Archives of General Psychiatry* 67, no. 3 (2010); A. Myers and J. C. Rosen, "Obesity Stigmatization and Coping: Relation to Mental Health Symptoms, Body Image, and Self-Esteem," *International Journal of Obesity and Related Metabolic Disorders* 23, no. 3 (1999); R. S. Strauss, "Childhood Obesity and Self-Esteem," *Pediatrics* 105, no. 1 (2000).

7   **After smoking, obesity is the leading cause:** Obesity is often touted as the "leading" cause of preventable death, even by experts. However, according to recent figures from the CDC, smoking still handily bests it 480,000 to 112,000 deaths per year.

7   **when it comes to morbidity:** Being diseased, on the other hand, is a lot more painful than being dead. See H. Jia and E. I. Lubetkin, "Trends in Quality-Adjusted Life-Years Lost Contributed by Smoking and Obesity," *American Journal of Preventative Medicine* 38, no. 2 (2010), and R. Sturm and K. B. Wells, "Does Obesity Contribute as Much to Morbidity as Poverty or Smoking?" *Public Health* 115, no. 3 (2001).

8   **The average American man has gained:** The average American man twenty to seventy-four years of age weighed 166.3 pounds in the 1960–1962 NEH survey, and the average male twenty and over weighed 195.5 pounds according to NHANES data for 2007–2010. The latter figure captures men older than seventy-four, and this would seem to bring the average down, because the group averages for men seventy and older are lower than the average of all men. The average woman in the 1960–1962 survey weighed 140.2 pounds and by 2007–2010 she weighed 166.2 pounds, as much as the average man fifty years earlier.

8   **The U.S. Weight Loss & Diet Control Market:** I am grateful to John LaRosa of Marketdata for supplying me with this information, which paints its own grim portrait. For example, the market for bariatric surgery was worth $350 million in 1989. By 2012, it had risen to $5.77 billion. Please note that from 1989 to 1996, the report was published every two years, and so for years with missing data I inserted data from whichever neighboring year was lower, erring on the side of conservatism. The yearly figures are not adjusted for inflation, so in current dollars the total is certainly much higher than $1 trillion.

8   **Fully one-third of boys and girls:** "Prevalence of Childhood and Adult Obesity in the United States, 2011–2012."

9   **includes twenty-four different substances:** Twenty-five, if you put arginine on the list, which is "conditionally essential"—premature babies can't make it, and certain medical situations, like surgery or sepsis, can place an undue demand on the body's ability to synthesize it.

11   **In the summer of 1962:** For the description of the West family trip from Dallas to Southern California during which Doritos were conceived, I am indebted to three of West's children: Greg West, Jack West, and Jana West. Frito-Lay declined to comment on, offer, or verify information.

12   **hit store shelves in 1964:** Other accounts place the debut of original Doritos in 1966 or 1967. However, Pepsico's UK website (www.pepsico.co.uk /doritos) both places the debut in 1964 and credits Arch West as the creator. This is significant. A competing Dorito narrative, as described in Gustavo Arellano's *Taco USA: How Mexican Food Conquered America*, claims their invention to have taken place at Disneyland. West's daughter, Jana, notes that Arch was friends with Walt Disney. There was also a restaurant at Disneyland called Casa de Fritos; therefore, the two accounts may not be irreconcilable. The original inspiration, however, according to every surviving member of the West family with whom I spoke—and Frito-Lay, too, it would seem—was the little Mexican "shack."

14   **an American corn farm was growing:** The yield figures are taken from the USDA's National Agricultural Statistics Service.

14   **but it tasted weaker:** My position that corn has gotten blander is based in part on a telephone interview with the pioneering and prolific USDA ARS flavor researcher Ron Buttery, who told me that "it's quite probable" that corn flavor changed as hybrid varieties and modern agricultural methods became more prevalent. He noted that open pollinated corn has more protein, and that less intensively raised corn (which is to say, corn as it was grown decades ago) has more phenolic compounds and vitamin C. Lastly, the enthusiasm among fine chefs and food lovers for heirloom varieties of corn—notably Eight Row Flint—further attests to this difference. See also the discussion of nutritional and flavor dilution in chapter 2. For perspective on the diminishment in potato flavor over decades, I am indebted to Mark Taylor, whose work I describe in chapter 9. I base my view also on having personally sampled Russet Burbank potatoes—a variety that was and still is popular for "chipping"—raised in an heirloom style. The tubers were bumpy and pitted but deeply potatoey and satisfying. I am also indebted to Shelley Jansky, whose chapter "Potato Flavor" in the *Handbook of Fruit and Vegetable Flavors* contains some interesting background on the effect of production environment on flavor, such as, "As nitrogen levels increase, sensory quality decreases, probably due to the production of acrid-tasting amides and amines."

16 **The thickest chapter:** According to Gordon Shepherd's *Neurogastronomy: How the Brain Creates Flavor and Why It Matters* (New York: Columbia University Press, 2012), the mammalian genome contains about thirty thousand genes, and 3 to 5 percent of them are for olfaction, or smell, constituting the "largest family in the genome." Since flavor combines both smell and taste—as we will see in chapter 3—the percentage of genes is even higher.

Two: Big Bland

19 **consider the following story:** This was shared with me in a telephone interview by heritage poultry farmer Frank Reese. Many of his elderly clientele relate similar Proustian poultry memories of their own from as far back as the Great Depression. As we will see in chapter 7, pastured heirloom chicken has an emotionally transportive effect even on people tasting it for the first time.

21 **began precisely sixty-one years earlier:** For so much of the history of early twentieth-century poultry, including the Chicken of Tomorrow contest, I am indebted to O. A. Hanke, *American Poultry History 1823–1973* (Madison, WI: American Poultry Historical Society, 1974), particularly its chapter 11, "Broilers."

21 **Americans almost doubled the amount of chicken:** According to disappearance data from the USDA's Economic Research Service, during World War II, annual per capita chicken consumption peaked at 22.4 pounds in 1945. In the five years prior to 1941, it had never exceeded 14.6 pounds.

25 **a five-pound chicken cost $3:** This is based on an annual average per pound price of 61¢ for ready-to-cook frying chickens in 1949, the earliest date I was able to find. The figure comes from the Bureau of Labor Statistics and I am grateful to the BLS's Darren Rippy for making the effort to locate it and taking the time to send it to me.

26 **"a 3 kg (6.6 lb) newborn baby":** R. F. Wideman et al., "Pulmonary Arterial Hypertension (Ascites Syndrome) in Broilers: A Review," *Poultry Science* 92, no. 1 (2013).

28 **Davis and his colleagues:** "Changes in USDA Food Composition Data for 43 Garden Crops, 1950 to 1999," *Journal of the American College of Nutrition* 23, no. 6 (2004). The data is laid out in column upon column of unsettling detail.

29 **Scientists had been aware of it:** For a deeper examination of the dilution effect that goes beyond Davis's 1999 paper, see his chapter, "Impact of Breeding and Yield on Fruit, Vegetable, and Grain Nutrient Content," in *Breeding for Fruit Quality* (New York: Wiley, 2011).

32 **Modern tomatoes are very, very bland:** The description of the biological underpinnings of blandness in tomatoes is based on numerous interviews,

in person and by phone, as well as e-mail correspondence with Harry Klee. Two of his published papers that cover this subject are "Genetic Challenges of Flavor Improvement in Tomato," *Trends in Genetics* 29, no. 4 (2013), and, with Linda Bartoshuk, "Better Fruits and Vegetables through Sensory Analysis," *Current Biology* 23, no. 9 (2013). The scientific account of how uniformly red tomatoes taste worse can be found in Ann L. T. Powell et al., "Uniform Ripening Encodes a Golden 2-like Transcription Factor Regulating Tomato Fruit Chloroplast Development," *Science* 336 (June 29, 2012).

33 **Modern broilers have been intentionally bred:** A recent paper in the *American Journal of Physiology—Endocrinology and Metabolism* found that fast-growing chickens are "resistant to the anorectic effect of exogenously administered CCK, suggesting that their satiety set point has been altered." In layman's terms, it takes them an awfully long time to get full. See also C. J. Rubin, "Whole-Genome Resequencing Reveals Loci Under Selection During Chicken Domestication," *Nature* 464 (March 25, 2010).

33 **For the first 99.9925 percent:** This is my own figure based on eight thousand years of domestication (an admittedly fuzzy figure) and sixty years of nutritionally fortified confinement feeding.

33 **their beaks would find the foodstuffs:** Here's a typical sentence from *The Complete Poultry Book: A Manual for the American Poultry Yard*, published in 1882: "If the range be large enough, this will be found not only the most economical plan, but that most conducive to the health of the fowls, as the exercise gained in hunting for food, and the variety of insect and green food thus obtained, will cause a thriftier growth than can be obtained by any artificial feeding."

34 **chickens got sick and died:** As John T. Henderson's *A Manual on Poultry*, published in 1883, put it, "They find on the natural range seeds of various kinds, a variety of green vegetable matter and insects, the three together supplying for them bread, vegetables and meat. If the birds are deprived of making their own selection of these classes of food in a natural manner, by reason of confinement within limited inclosures, they must be supplied by artificial means or the fowls will suffer from the privation, and be unprofitable to their owners."

34 **It was, in fact, vitamin B$_1$:** The story of how vitamins and their varied and essential nature was finally achieved is, in fact, a good deal more complicated. But to keep it simple, Eijkman thought white rice contained a poison that caused beriberi and the bran contained the antidote. Funk took it a crucial step farther and formed the idea that the missing substance wasn't a cure but an essential nutrient. He coined the now ubiquitous and basic concept of vitamins. Thiamin, however, wasn't isolated until 1926.

35 **it's the same reason a dairy cow:** The best-known example of this in poultry is that chickens that are fed fish taste fishy (and not in a good way). For

a general overview, see Leo M. Nollet, ed., *Handbook of Meat, Poultry and Seafood Quality* (New York: Wiley-Blackwell, 2012).

36 **learned this lesson as early as 1917:** The knowledge that there is a relationship between a bird's age and its flavor hasn't been lost so much as ignored. The 1997 edition *Joy of Cooking* has this to say about chicken: "In order to remain profitable in a highly competitive industry, producers rush their chickens to market less than 40 days after hatching, which is simply not enough time for the birds to develop flavor."

39 **nearly half of all chicken sold is "further processed":** According to Tom Super, vice president of communications at the National Chicken Council, who told me by e-mail correspondence that 40 percent of chicken sold to "retail and food service outlets" is further processed.

Three: Big Flavor

41 **as glamorous as spying:** Kaestner did some spying, too. Following his trips to Madagascar in the mid-1970s, he would be debriefed in his office by a contact from the CIA. One year, Kaestner snapped some photos of Soviet MiG planes under camouflaged netting at the main airport in Antananarivo. This was, he says, considered a particularly valuable piece of intelligence.

42 **By 1976, Madagascar was producing:** Madagascar's vanilla production figures were taken from "Economic Prospects for Vanilla in the South Pacific," published by the Australian Centre for International Agricultural Research.

44 **"silky-sheened, very delicate":** The story of Wilhelm Haarmann's quest for artificial vanillin comes from Björn Bernhard Kuhse, *Wilhelm Haarmann auf den Spuren der Vanille*, 2012. I am grateful to my mother, Valerie Schatzker, for translating key passages of the book. I am grateful also to Bernhard Kott of Symrise AG for sending it to me.

44 **a fundamentally different kind of smelling:** The inner workings of flavor as described here are based on Gordon Shepherd's *Neurogastronomy*, as well as numerous telephone and in-person interviews with Linda Bartoshuk. Barb Stuckey, *Taste What You're Missing* (New York: Simon & Schuster, 2012), also has a good description.

45 **distinguishing more than a trillion aromas:** For decades, scientists thought the total number was more like ten thousand. This new and much, much larger number comes from C. Bushdid et al., "Humans Can Discriminate More than 1 Trillion Olfactory Stimuli," *Science* 343 (March 21, 2014) which, as the paper neatly puts it, "demonstrates that the human olfactory system, with its hundreds of different olfactory receptors, far outperforms the other senses in the number of physically different stimuli it can discriminate."

46 **taste only a very mild bitter flavor:** Telephone interview with Marianne Gillette.

47 **Vanilla contains hundreds of other aromatic compounds:** A. K. Sinha, U. K. Sharma, and N. Sharma, "A Comprehensive Review on Vanilla Flavor: Extraction, Isolation and Quantification of Vanillin and Other Constituents," *International Journal of Food Sciences and Nutrition* 59, no. 4 (2008).

50 **In 1965, there were less than 700:** I calculated this number personally (but in consultation with Mat Gulick, representing the Flavor and Extract Manufacturers Association) from the first published list of Generally Regarded as Safe flavoring substances, in GRAS III, *Food Technology* 19. no. 2 (1965). I arrived at my total by subtracting the number of whole ingredients or extracts with "species names," such as dill seed and dill oil, from the total, leaving me with the approximate number of pure chemical constituent flavor compounds, composed of substances such as 2-Dodecenal and methyl acetate.

50 **there are more than 2,200:** According to correspondence with the Flavor and Extract Manufacturers Association, the GRAS list today stands at more than twenty-seven hundred different substances. I subtracted the same number of ingredients with species names from the total, as with GRAS III, to arrive at more than twenty-two hundred.

50 **The most recent additions:** I am grateful to FEMA's Mat Gulick for providing me with these recent additions to the GRAS list.

50 **produce knockoffs:** From the Givaudan SA 2013 annual report: "Givaudan has new Expressive Naturals Collections in three families—citrus, apple and assorted—and each list highlights some of the newest and most novel additions to the portfolio. Infused with named origin ingredients, these authentic expressions of nature's finest fruits allow customers to connect the imagery and message of their product with the true-to-source flavour profile. And the identification of named fruits and sources engages consumers looking to retain a sense of local identity and speciality. Apple offerings, for example, include the signature apple orchards of America with favourites such as Honeycrisp and Golden Delicious, while the assorted list includes the new world of sloe berry and jasmine green tea."

50 **soon delta-dodecalactone:** History departments should collectively bow their heads in shame over the totally inadequate effort to document the history of flavor technology. For so much of my description, I owe a debt of gratitude to John Leffingwell (www.leffingwell.com), an industry veteran who remains an invaluable source of information on this important but ignored area. In particular, I am grateful for his PowerPoint presentation "50 Years of Flavor Chemistry," the facts of which were confirmed by telephone interview.

51 **By the late 1960s, the umami arsenal:** Once again, where the history

departments fail, John Leffingwell succeeds. In a telephone interview, he related the use of these ingredients in the food industry in the late 1960s.

51 **In the early 1990s, scientists at Ajinomoto:** So far as I can tell, Ajinomoto doesn't employ a single media relations specialist, or if it does, that person wishes not to communicate with me. I am grateful to Ajinomoto's Joe Formanek for his insights into the development and use of kokumi ingredients. For scientific publications on kokumi, see Y. Ueda et al., "Flavor Characteristics of Glutathione in Raw and Cooked Foodstuffs," *Bioscience, Biotechnology, and Biochemistry* 61, no. 12 (1997), and T. Ohsu et al., "Involvement of the Calcium-Sensing Receptor in Human Taste Perception," *Journal of Biological Chemistry* 285, no. 2 (2010).

59 **The actual molecules sitting in the beaker**: To be clear, many artificial flavor chemicals have never been found in nature, even though they smell very much like something natural. But many, many artificial flavors are simply identical man-made versions of chemicals produced in nature.

61 **It is an astonishing increase:** According to the USDA's "Nutrient Content of the US Food Supply, 1909–2000," "The use of spices increased more than fivefold from one-half pound per capita in 1918 to 2.59 pounds per capita in 2000." The USDA's ERS has slightly different numbers (because it uses different population estimates), starting at 0.8 pound per person in 1918 and reaching 3.5 pounds per person in 2010, which, at 438 percent, is very close.

61 **The industry was in its infancy:** For this admittedly brief portrait of the early twentieth-century flavor industry, I am indebted to John Leffingwell as well as *FEMA 100: A Century of Great Taste.*

62 **Smithfield's pork shoulder roast:** Smithfield's Foods' Kathleen Kirkham did not respond to repeated e-mails and telephone messages to confirm these ingredients.

62 **"beef and natural flavorings":** These ingredients are listed on Wal-Mart's online grocery store for its fresh rib eye steak, although similar ingredients can be found on the packaging of beef by other purveyors. Wal-Mart's top sirloin and boneless round steak also feature natural flavorings.

63 **Flavor chemicals of one sort or another:** The defense will invariably be made that some of these "natural flavors" are present in tiny quantities and present for some totally innocent reason, such as to preserve a flavor lost during canning, or boost the flavor of chopped tomatoes during a poor harvest. Such uses, however innocent their intended effect, may be smoothing over nutritional deficiencies the palate is normally attuned to. More important, unless an ingredients list states exactly which substances are in a product, the onus is on the maker to prove their authenticity and not on the consumer to assume it.

63 **You can even find curry sauce:** Maya Kaimal Madras Curry Indian Simmering Sauce, a Williams-Sonoma "exclusive," features natural flavor and

hydrolyzed yeast. Luvo's Thai Style Green Chicken Curry with Green Tea Infused Brown Rice & Farro is certainly one of the most wholesome-*sounding* offerings in the curry category, but it would appear to be among the most chemically adulterated. The ingredients include four separate instances of "natural" flavor along with yeast extract.

64 **"not to club them senseless":** Marcella Hazan, *Marcella Says . . .* (New York: William Morrow, 2004), page 54. Herewith, a few other choice quotes from a woman who possessed a deep and wise understanding of flavor: "I don't have a spice rack in my kitchen, nor do I believe you will find many Italian kitchens that do," "the unbalanced use of garlic is the single greatest cause of failure in the would-be Italian cooking," flavor "wants to be disclosed rather than imposed," and flavor "is not the idea of a thing, but the thing itself."

64 **adding flavor chemicals to actual butter:** The following is by no means an exhaustive list of companies that add natural flavorings to butter: Land O'Lakes, Challenge, Plugra, and Tillamook. Patricia from Land O'Lakes consumers affairs told me, via e-mail, that the natural flavoring in Land O'Lakes Unsalted Butter is lactic acid. This is most likely intended to simulate the flavor of cultured butter, which also contains lactic acid. However, the flavoring is deceptive. There is more than just lactic acid separating cultured butter from sweet cream butter. Cultured butter also features less lactose than uncultured butter, as well as bacterial culture, which may contribute to digestive health.

65 **while we're on the subject of California:** California figures are based on the USDA's National Agricultural Statistics Service historical state data. The national data that follow is also from the NASS.

65 **Hens lay twice as many eggs:** I am indebted to the food historian Charlie Gracey for providing me with information on efficiency improvements in livestock.

66 **is worth $2 billion:** There is no generally agreed upon value of the U.S. or global flavor industry, in part due to the fact that measuring it is difficult. Some companies, such as large soft drink makers, manufacture their own flavorings, and the value of these flavorings is difficult to capture since they are not reported as individual sales.

66 **Givaudan has twenty flavor factories:** This number is derived from Givaudan's 2013 annual report by counting the number of Givaudan global legal entities that participate in both "flavours" and "production." Givaudan, when contacted, would not reveal the exact number of its flavor factories, but Peter Wullschleger, of Givaudan Media and Investor Relations, suggested I count them in the annual report.

66 **more than 1.4 million tons:** Courtesy of Euromonitor International.

Four: Big People

67 **And the most terrible cookies of all:** Nidetch does not name Mallomars as the cookie she secretly gorged on in *The Story of Weight Watchers*, but she does say "that [Mallomars] was my Frankenstein," in Nanci Hellmich, "Jean Nidetch Shares Her Story," *USA Today*, March 23, 2010.

67 **a research team:** I am grateful to Sonja Yokum of the Oregon Research Institute for providing me with details of both the test and the subjects so that I could write this portrait.

70 **craves food more than skinny Sarah:** See also E. Stice et al., "Relation of Reward from Food Intake and Anticipated Food Intake to Obesity: A Functional Magnetic Resonance Imaging Study," *Journal of Abnormal Psychology* 117, no. 4 (2008).

71 **ten white rats lived the all-you-can-eat life:** This study, "Dopamine D2 Receptors in Addiction-like Reward Dysfunction and Compulsive Eating in Obese Rats," *Nature Neuroscience* 13 (2010) was the PhD thesis of Paul M. Johnson. I am grateful to him for taking the time to share extra details and explain the context of his findings as they relate to the existing literature on human addiction and the modern food environment. For more of the scientific literature on addiction, see his bibliography.

72 **intracranial self-stimulation:** The electrodes used in intracranial self-stimulation allow a rat to stimulate a reward region of its brain by pressing a lever. Simply put, the more the rat presses the lever, the less reward the rat is deriving from its environment. In this study, the rats gorging on food pressed the lever rarely at the beginning of the study but more and more as the study went on, as their BMI got higher and higher.

73 **Salt, sugar, and fat are what psychologists call reinforcers:** For an excellent and highly readable discussion of the reinforcing nature of salt, sugar, and fat, see David A. Kessler, *The End of Overeating: Taking Control of the Insatiable American Appetite* (New York: Rodale, 2009).

73 **The food companies know this:** For a chilling portrait of exactly how food companies have used salt, sugar, and fat to bolster profits, see Michael Moss, *Salt Sugar Fat* (New York: Random House, 2014).

73 **umami stimulates the orbitofrontal cortex:** From I. E. Araujo, "Representation of Umami Taste in the Human Brain," *Journal of Neurophysiology* 90, no. 1 (2003): "The evidence described here that the orbitofrontal cortex represents umami taste (literally "delicious" in Japanese) is consistent with other evidence that the orbitofrontal cortex also represents information about the reward value of other primary and secondary reinforcers including taste." See also E. T. Rolls, "The Representation of Umami Taste in the Taste Cortex," *Journal of Nutrition* 130 (1999), and Y. Nakamura et al., "Localization of Brain Activation by Umami Taste in Humans," *Brain Research* 1406 (2011).

74 **people become impulsive:** For a discussion of the neuroscience relating OFC dysfunction to cocaine addiction and poor decision making, see Federica Lucantonio et al., "The Impact of Orbitofrontal Dysfunction on Cocaine Addiction," *Nature Neuroscience* 15 (2012).

74 **women who answered an ad offering $350:** The story is told in B. Spring et al., "Abuse Potential of Carbohydrates for Overweight Carbohydrate Cravers," *Psychopharmacology* 197, no. 4 (2008).

74 **The carb-craving overweight women loved the high-carb drink:** The actual drink was Carbo Force, and the carbohydrates contained within were maltodextrin, dextrose, high maltose rice syrup, palantinose, rice oligodextrine, and D-ribose. While some of these ingredients are perceived as sweet, their sweetness is much lower, relatively speaking, than sugar, underscored by the fact that Carbo Force also contains the artificial sweetener sucralose. This further suggests a strong component of the drink's effect on subjects was "post-ingestive," supported by the extended period of elevated mood after drinking it.

75 **subtler aspects of addiction:** I am indebted to Serge Ahmed, of the Institut des Maladies Neurodégénératives at the University of Bordeaux for walking me through the epidemiology of addiction and clarifying my understanding of two variables that drive numbers of addicts: potency and availability.

75 **Not everyone who tries addictive drugs:** James C. Anthony, "Epidemiology of Drug Dependence," in *Neuropsychopharmacology: The Fifth Generation of Progress*, ed. Kenneth L. Davis et al. (Philadelphia: Lippincott, 2002).

76 **an astonishing number of American Army soldiers:** See Lee N. Robins's fascinating accounts, "Vietnam Veterans' Rapid Recovery from Heroin Addiction: A Fluke or Normal Expectation?" *Addiction* 88, no. 8 (1993), and "Drug Use by U.S. Army Enlisted Men in Vietnam: A Follow-Up on Their Return Home," *American Journal of Epidemiology* 99, no. 4 (1974).

79 **As an industry newsletter puts it:** "Research News No. 16: Sucram," published by Pancosma.

80 **an improved weight gain of 30 percent:** "Molasweet Palatant Boosts Lamb Growth," May 27, 2008, www.allaboutfeed.net.

84 **Only around 10 percent of Italian adults are obese:** "Obesity and the Economics of Prevention: Fit not Fat," OECD, 2010, www.oecd.org/els/health-systems/49712603.pdf.

Five: The Wisdom of Flavor

92 **But here's what was convincing:** The description of how Joe Lewis and Jim Tumlinson and their colleagues came to understand the complex signaling relationship among plants, herbivorous insects, and parasitic wasps took place over many years and was reported in numerous scientific papers.

My portrait is based primarily on phone and e-mail interviews with both scientists. Perhaps the best overview is their "How Parasitic Wasps Find Their Hosts," *Scientific American*, March 1993. The scientific references include T. C. J. Turlings, J. H. Tumlinson, and W. J. Lewis, "Exploitation of Herbivore-Induced Plant Odors by Host-Seeking Parasitic Wasps," *Science* 250 (November 30, 1990); T. C. Turlings and J. H. Tumlinson, "Systemic Release of Chemical Signals by Herbivore-Injured Corn," *Proceedings of the National Academy of Sciences* 89, no. 17 (1992); and T. C. Turlings et al., "How Caterpillar-Damaged Plants Protect Themselves by Attracting Parasitic Wasps," *Proceedings of the National Academy of Sciences* 92, no. 10 (1995).

94 **It must be learned:** F. D. Provenza et al., "Conditioned Flavor Aversion: A Mechanism for Goats to Avoid Condensed Tannins in Blackbrush," *American Naturalist* 136 (1990).

95 **a German pharmacist isolated morphine:** An excellent primer on the history of the science of plant secondary metabolism is Thomas Hartmann, "From Waste Products to Ecochemicals: Fifty Years' Research of Secondary Metabolism," *Phytochemistry* 66 (2007)

96 **to ward off hungry insects:** Gottfried Fraenkel's amazing theory, which gave birth to the field known as chemical ecology, is as fresh and startling to read now as the day it was published in 1959. See "The Raison d'Être of Secondary Plant Substances," *Science* 129 (May 29, 1959). I am grateful to the renowned entomologist May Berenbaum of the University of Illinois for guiding my understanding of this subject area (and tolerating my questions). Her paper, "Facing the Future of Plant-Insect Interaction Research: Le Retour à la 'Raison d'Être,'" *Plant Physiology* 146, no. 3 (2008), provides an excellent overview of Fraenkel's argument, a snapshot of how the field has matured, and even more insight into the complex and constantly advancing nature of chemical "warfare" between species.

97 **Provenza looked at the mineral phosphorus:** "Phosphorus Appetite in Sheep: Dissociating Taste from Postingestive Effects," *Journal of Animal Science* 84, no. 8 (2006). Note that in this and other papers by Provenza I have excluded certain experimental details for the sake of clarity. For example, this experiment also included ruminal infusions of salt, to differentiate phosphorus liking from salt liking, as well as a third flavoring, onion. Crucially, Provenza also controlled for "novelty" by exposing each group of sheep to all three flavor-nutrient interactions simultaneously. The sheep were not simply choosing any diet other than the deficient one. I related my telling of it and other experiments to Fred Provenza, who felt it was accurate.

98 **Provenza did a similar experiment with calcium:** J. J. Villalba, F. D. Provenza, and J. O. Hall, "Learned Appetites for Calcium, Phosphorus, and Sodium in Sheep," *Journal of Animal Science* 86, no. 3 (2008).

98   **a stomach drenching of carbs:** Juan J. Villalba and Frederick D. Provenza, "Postingestive Feedback from Starch Influences the Ingestive Behaviour of Sheep Consuming Wheat Straw," *Applied Animal Behavior Science* 66 (2000). In this study, flavors were not added to straw. Rather, two different groups of sheep ate straw, but one group received a bigger ruminal infusion of starch. Therefore, the flavor of the straw itself was paired with different post-ingestive consequences for each group, and this had an effect on intake.

98   **Provenza showed that it happened with protein:** J. J. Villalba and F. D. Provenza, "Nutrient-Specific Preferences by Lambs Conditioned with Intraruminal Infusions of Starch, Casein, and Water," *Journal of Animal Science* 77, no. 2 (1999). This study is particularly interesting because it shows sheep discriminating post-ingestive "feedbacks" from both energy and protein and balancing each within a single meal.

98   **formed a preference for more protein:** I. Kyriazakis and J. D. Oldham, "Diet Selection in Sheep: The Ability of Growing Lambs to Select a Diet That Meets Their Crude Protein (Nitrogen × 6.25) Requirements," *British Journal of Nutrition* 69, no. 3 (1993).

99   **this time with calves:** S. B. Atwood et al., "Influence of Free-Choice vs. Mixed-Ration Diets on Food Intake and Performance of Fattening Calves," *Journal of Animal Science* 79, no. 12 (2001).

100   **cut an opening in the throats of six goats:** F. D. Provenza and J. C. Malechek, "Diet Selection by Domestic Goats in Relation to Blackbrush Twig Chemistry," *Journal of Applied Ecology* 21 (1984).

101   **The substance is oxygen:** Barry Halliwell, "Polyphenols: Antioxidant Treats for Healthy Living or Covert Toxins?" *Journal of the Science of Food and Agriculture* 86, no. 13 (2006), features an excellent discussion of oxygen's transformation from toxin to essential substance.

102   **they seemed to cause more cancer:** "CARET was stopped ahead of schedule in January 1996 because participants who were randomly assigned to receive the active intervention were found to have a 28% increase in incidence of lung cancer, a 17% increase in incidence of death and a higher rate of cardiovascular disease mortality compared with participants in the placebo group." From G. E. Goodman et al., "The Beta-Carotene and Retinol Efficacy Trial: Incidence of Lung Cancer and Cardiovascular Disease Mortality During 6-Year Follow-up After Stopping β-Carotene and Retinol Supplements," *Journal of the National Cancer Institute* 96, no. 23 (2004).

104   **much less toxic when eaten together:** Travis E. Mote, Juan Villalba, and Frederick D. Provenza, "Sequence of Food Presentation Influences Intake of Foods Containing Tannins and Terpenes," *Applied Animal Behavior Science* 113 (2008), and Steven S. Seefeldt, "Consequences of Selecting Rambouillet Ewes for Mountain Big Sagebrush (*Artemisia tridentata* ssp.

vaseyana) Dietary Preference," *Rangeland Ecology & Management* 58, no. 4 (2005).

104 **tannin consumption shot up:** "Preference for Condensed Tannins by Sheep in Response to Challenge Infection with *Haemonchus Contortus*," *Veterinary Parasitology* 188 (2012).

104 **When tiger moth caterpillars were infected:** Elizabeth A. Bernays and Michael S. Singer, "Insect Defences: Taste Alteration and Endoparasites," *Nature*, July 28, 2005.

104 **flavonoids gave their immune system a boost:** C. Catoni, H. Martin Schaefer, and A. Peters, "Fruit for Health: The Effect of Flavonoids on Humoral Immune Response and Food Selection in a Frugivorous Bird," *Functional Ecology* 22, no. 4 (2008).

104 **Honeybees respond to a fungal infection:** Michael D. Simone-Finstrom and Marla Spivak, "Increased Resin Collection after Parasite Challenge: A Case of Self-Medication in Honey Bees?" *PLOS ONE*, March 29, 2012.

104 **chimpanzees chewed a plant called bitter leaf:** Michael A. Huffman, "Self-Medicative Behaviour in the African Great Apes: An Evolutionary Perspective into the Origins of Human Traditional Medicine," *BioScience* 51, no. 8 (2001), and Michael A. Huffman et al., "Further Observations on the Use of the Medicinal Plant, *Vernonia amygdalina* (Del), by a Wild Chimpanzee, Its Possible Effect on Parasite Load, and Its Phytochemistry," *African Study Monographs* 14 (1993).

105 **sick sheep preferred the flavor:** J. J. Villalba, F. D. Provenza, and R. Shaw, "Sheep Self-Medicate When Challenged with Illness-Inducing Foods," *Animal Behaviour* 71, no. 5 (2006).

107 **that will bring its fetus protein:** S. D. B. Cooper, I. Kyriazakis, and J. D. Oldham, "The Effect of Late Pregnancy on the Diet Selections Made by Ewes." *Livestock Production Science* 40, no. 3 (1994).

Six: Bait and Switch

109 **merlot became famously untrendy:** Steven Cuellar, Dan Karnowsky, and Frederick Acosta, "The 'Sideways' Effect: A Test for Changes in the Demand for Merlot and Pinot Noir Wines," American Association of Wine Economists, October 2008, www.wine-economics.org, notes, interestingly, that the positive effects on pinot noir were greater than the negative effects on merlot.

110 **lit up like suburban Christmas lights:** "Our results show that increasing the price of a wine increases subjective reports of flavor pleasantness as well as blood-oxygen-level-dependent activity in medial orbitofrontal cortex," in Hilke Plassmann et al., "Marketing Actions Can Modulate Neural Representations of Experienced Pleasantness," *Proceedings of the National Academy of Sciences* 105, no. 3 (2008).

110 **We *like* the wrong food:** I am grateful to Anthony Sclafani, distinguished professor of psychology and director of the Feeding Behavior and Nutrition Laboratory at Brooklyn College, for taking the time to familiarize me with the literature on flavor preference in humans, which is characterized by statements such as "flavours not only become liked by pairing with energy but also become cues that engage wanting for foods, in particular palatable (often high sugar, high fat) foods, since these are both common and effective sources of energy," from J. Prescott, "Chemosensory Learning and Flavour: Perception, Preference and Intake," *Physiology and Behaviour* 107, no. 4 (2012). See also Martin R. Yeomans, "Development of Human Learned Flavor Likes and Dislikes," in *Obesity Prevention: The Role of Brain and Society on Individual Behavior,* ed. Laurette Dubé et al. (Burlington, MA: Elsevier, 2010); M. R. Yeomans, "Flavour-Nutrient Learning in Humans: An Elusive Phenomenon?" *Physiology and Behaviour* 106, no. 3 (2012); Stephanie Stein et al., "Food Intake and Food Preference," in *Diet Quality: An Evidence-Based Approach, Volume 1,* ed. Victor R. Preedy, La-Anh Hunter, and Vinood B. Patel, e-book, link.springer.com; P. A. Breslin, "An Evolutionary Perspective on Food and Human Taste," *Current Biology* 23, no. 9 (2013).

110 **"We eat because":** Roger Cohen, "Fat Britain," *New York Times,* July 10, 2014.

111 **began craving the one substance:** See Vilhjalmur Stefansson, *The Fat of the Land* (New York: Macmillan 1956), in which the arctic explorer discusses the dangers of a lean meat diet in the Arctic and his own desire for fattier meat. See also *The Prairie Traveler: A Hand-book for Overland Expeditions* (1859), in which Randolph B. Marcy writes, "We tried the meat of horse, colt, and mules, all of which were in a starved condition, and of course not very tender, juicy, or nutritious. We consumed the enormous amount of from five to six pounds of this meat per man daily, but continued to grow weak and thin, until, at the expiration of twelve days, we were able to perform but little labor, and were continually craving for fat meat." Lastly, John D. Speth and Katherine A. Spielmann, "Energy Source, Protein Metabolism, and Hunter-Gatherer Subsistence Strategies," *Journal of Anthropological Archeology* 2, no. 1 (1983). http://en.wikipedia.org/wiki/Vilhjalmur_Stefansson.

111 **Pregnant women in the tropics:** For an excellent review of geophagy along with one of the best academic examinations of preference of foods higher in plant secondary compounds, see Timothy Johns, *The Origins of Human Diet and Medicine* (Tucson: University of Arizona Press, 1996).

112 **persuaded several teenage mothers and widows:** Clara M. Davis, "Results of the Self-Selection of Diets by Young Children," *Canadian Medical Association Journal* 41, no. 13 (1939).

113 **color and aroma compounds:** For more on the delectability of ripe fruit,

see Gordon Shepherd's discussion of fruit aromas in chapter 4 of *Neurogastronomy*.

114 **a density of secondary compounds:** John A. Manthey and Penelope Perkins-Veazie, "Levels of β-Carotene, Ascorbic Acid, and Total Phenols in the Pulp of Five Commercial Varieties of Mango (*Mangifera indica* L.)," *Proceedings of the Florida State Horticultural Society* 122 (2009).

114 **similarly pack more secondary compounds:** Andrea Bunea et al., "Comparative Polyphenolic Content and Antioxidant Activities of Some Wild and Cultivated Blueberries from Romania," *Notulae Botanicae Horti Agrobotanici Cluj-Napoca* 39, no. 2 (2011).

115 **It used to be served in Nevada brothels:** Jeremy Agnew, *Medicine in the Old West: A History: 1850–1900* (Jefferson, NC: McFarland, 2010).

115 **at least six thousand years:** Hayley Saul et al., "Phytoliths in Pottery Reveal the Use of Spice in European Prehistoric Cuisine," *PLOS ONE*, August 21, 2013.

115 **found in the teeth of Neanderthals:** K. Hardy et al., "Neanderthal Medics? Evidence for Food, Cooking, and Medicinal Plants Entrapped in Dental Calculus," *Naturwissenschaften* 99, no. 8 (2012).

116 **"inhibits pro-inflammatory mediator expression":** T. T. Wu et al., "Suppressive Effects of Extracts from the Aerial Part of *Coriandrum Sativum* L. on LPS-Induced Inflammatory Responses in Murine RAW 264.7 Macrophages," *Journal of the Science of Food and Agriculture* 90, no. 11 (2010).

116 **Fennel extract exhibits:** Weiping He and Baokang Huang, "A Review of Chemistry and Bioactivities of a Medicinal Spice: *Foeniculum vulgare*," *Journal of Medicinal Plants Research* 5, no. 16 (2011).

116 **Ginger alleviates:** *Ginger*, McCormick & Company, 2009.

116 **Dill promotes skin elasticity:** B. Sohm et al., "Evaluation of the Efficacy of a Dill Extract in Vitro and in Vivo," *International Journal of Cosmetic Science* 33, no. 2 (2011).

116 **Basil kills viruses:** *Basil*, McCormick & Company, 2013.

116 **cinnamon decreases blood glucose:** Keith Singletary, "Cinnamon: Overview of Health Benefits," *Nutrition Today* 43, no. 6 (2008).

116 **black pepper exhibits antidepressant:** *Black Pepper*, McCormick & Company, 2009.

116 **Cloves modify platelet activity:** R. H. Raghavendra and K. A. Naidu, "Spice Active Principles as the Inhibitors of Human Platelet Aggregation and Thromboxane Biosynthesis," *Prostaglandins, Leukotrienes and Essential Fatty Acids* 8, no. 1 (2009).

116 **elderly Singaporeans who eat curry:** "Curry Consumption and Cognitive Function in the Elderly," *American Journal of Epidemiology*.

116 **It is also thought that turmeric:** Keith Singletary, "Turmeric: An Overview of Potential Health Benefits," *Nutrition Today* 45, no. 5 (2010).

117 **"shiverings," "tremblings," and "putrid fever":** Richard Walter, *A Voyage*

Round the World in the Years 1740–1744, by George Anson, Esq., Compiled from His Papers and Materials by Richard Walter, Chaplain to the Centurion in That Expedition, 1804, p. 103.

117   **forty-three men had died at sea:** Ibid., p. 102.

117   **Theories ran the gamut:** E-mail from Jonathan Lamb, historian of voyaging at Vanderbilt University and author of *Preserving the Self in the South Seas, 1680–1840.*

117   **"In our distressed situation":** *A Voyage round the World,* p. 111.

118   **found watercress and purslane:** Ibid., p. 118.

118   **"gather strength even from the sight of the fruit":** Thomas Trotter, *Observations on the Scurvy,* 1792, p. 142.

118   **"the spirits are exhilarated":** Ibid.

118   **first clinical trial:** Arun Bhatt, "Evolution of Clinical Research: A History Before and Beyond James Lind," *Perspectives in Clinical Research* 1, no. 1 (2010).

118   **"Nature points out the remedy":** James Lind, *On the Causes of Scurvy Part II,* p. 83.

118   **"a strange plethora":** Jonathan Lamb, "Captain Cook and the Scourge of Scurvy," www.bbc.co.uk/history/british/empire_seapower/captaincook _scurvy_01.shtml.

119   **famous health benefits of the Mediterranean Diet:** Nutritionists may not agree on much, but the epidemiological track record of the Mediterranean Diet is as solid as epidemiological data gets. See T. N. Sergentanis et al., "Mediterranean Diet, Stroke, Cognitive Impairment, and Depression: A Meta-Analysis," *Annals of Neurology* 74, no. 4 (2013); E. Koloverou, "The Effect of Mediterranean Diet on the Development of Type 2 Diabetes Mellitus: A Meta-Analysis of 10 Prospective Studies and 136,846 Participants," *Metabolism* 63, no. 7 (2014); and Ramón Estruch et al., "Primary Prevention of Cardiovascular Disease with a Mediterranean Diet," *New England Journal of Medicine* 368 (2013).

122   **We have similar chemical sensors:** For those who contend that humans are lesser smellers because of the high number of olfactory pseudogenes, have a look at chapter 26 of Gordon Shepherd's *Neurogastronomy,* which points out that pseudogene-encumbered primates, including humans, nevertheless possess excellent senses of smell.

123   **A study at the University of Liverpool:** A. W. Moran et al., "Expression of Na+/glucose Co-Transporter 1 (SGLT1) Is Enhanced by Supplementation of the Diet of Weaning Piglets with Artificial Sweeteners," *British Journal of Nutrition* 104, no. 5 (2010).

123   **artificial sweeteners don't seem to be doing us much good:** S. E. Swithers, "Artificial Sweeteners Produce the Counterintuitive Effect of Inducing Metabolic Derangements," *Trends in Endocrinology and Metabolism* 24, no. 9 (2013).

124 **Dana Small hypothesized:** K. J. Rudenga and D. M. Small, "Amygdala Response to Sucrose Consumption Is Inversely Related to Artificial Sweetener Use," *Appetite* 58. no. 2 (2012).

124 **A recent job posting by Philip Morris International:** www.qual.ch /permlinks_job_2863.html?lang=fr.

125 **a master list of all the ingredients:** www.rjrt.com/tobaccoingredients .aspx. Phillip Morris USA publishes a similar list, www.philipmorrisusa .com/en/cms/Products/Cigarettes/Ingredients/Tobacco_Flavor_Ingredients /default.aspx?src=top_nav.

125 **"The taste of licorice":** "Tobacco Flavoring for Smoking Products," R. J. Reynolds Tobacco Company.

125 **"make the product sell better":** Robert B. Felton, "What Flavoring Can Do to Improve Sales of Tobacco Products," Felton International Inc., 1972, obtained, gratefully, at the Legacy Tobacco Documents Library, legacy .library.ucsf.edu/tid/xtv82e00/pdf;jsessionid=274843D9E21B8CE3D53A 27CB2A85A27E.tobacco03.

126 **"We get a very good idea what children consider":** Peter Wullschleger, Givaudan Media and Investor Relations, telephone interview, March 12, 2014.

126 **would eat even more hay:** D. M. Early and F. D. Provenza, "Food Flavor and Nutritional Characteristics Alter Dynamics of Food Preference in Lambs," *Journal of Animal Science* 76, no. 3 (1998).

126 **The same effect has been documented numerous times with rats:** M. Naim et al., "Energy Intake, Weight Gain and Fat Deposition in Rats Fed Flavored, Nutritionally Controlled Diets in a Multichoice ('Cafeteria') Design," *Journal of Nutrition* 115, no. 11 (1985), and D. Treit, M. L. Spetch, and J. A. Deutsch, "Variety in the Flavor of Food Enhances Eating in the Rat: A Controlled Demonstration," *Physiology and Behavior* 30, no. 2 (1983).

126 **scientists in Italy sprayed ryegrass:** R. G. De et al., "Influence of Flavor on Goat Feeding Preferences," *Journal of Chemical Ecology* 28, no. 2 (2002).

126 **Scientists at Japan's National Grassland Research Institute:** H. Dohi, A. Yamada, and T. Fukukawa, "Intake Stimulants in Perennial Ryegrass (*Lolium perenne* L.) Fed to Sheep," *Journal of Dairy Science* 80, no. 9 (1997), and "Effects of Organic Solvent Extracts from Herbage on Feeding Behavior in Goats" *Journal of Chemical Ecology* 22, no. 3 (1996).

132 **one of nature's greatest affronts to vegetarianism:** Dawn R. Bazely, "Carnivorous Herbivores: Mineral Nutrition and the Balanced Diet," *Trends in Ecology & Evolution* 4, no. 6 (1989).

132 **caribou on the shores of Hudson Bay:** As witnessed by Dawn Bazely, professor of biology at York University, related to the author in a telephone interview.

132 **Since folic acid enrichment was introduced:** "The Ups and Downs of Folic

Acid Fortification," *Harvard Women's Health Watch*, 2008, www.health
.harvard.edu/newsweek/the-ups-and-downs-of-folic-acid-fortification
.htm.

133 **a whole bunch of "enriched" nutrients:** For the enriched nutritional
makeup of Froot Loops with Fruity Shaped Marshmallows, I am indebted
to Donald Davis, who pointed out, via e-mail, that "the full Nutrition Facts
label shows 25% of the DV in a 110-Calorie serving (30% with added
milk). This is much more than would be normally present in 110 Calories
of grain." A Kellogg's employee at the media hotline refused to provide
me with information on enrichment unless I provided Kellogg's with the
full chapter to read first. Besides being absurd—how can I write a chapter
before I've researched it?—this request is journalistically unethical. Kellogg's did not respond to a detailed e-mail request.

Seven: Fried Chicken Saved My Life!

143 **all 1,120 calories:** Calories calculated at www.mcdonalds.ca/ca/en/food
/nutrition_centre.html#/.

144 **Back in 1954:** M. D. Sweetman and I. MacKellar, Food Selection and Prep-
aration (New York: Wiley, 1954); also H. S. Ramaswamy and J. F. Richards,
"Flavor of Poultry Meat—A Review," *Canadian Institute of Food Science
Technology Journal* 15 (1982).

145 **In 1997, Farmer did a study:** L. J. Farmer et al., "Responses of Two Genotypes of Chicken to the Diets and Stocking Densities of Conventional UK
and Label Rouge Production Systems—II. Sensory Attributes," *Meat Science* 47 (1997).

145 **compared chickens of today:** Y. Wang et al., "Modern Organic and Broiler
Chickens Sold for Human Consumption Provide More Energy from Fat
than Protein," *Public Health Nutrition* 13, no. 3 (2010).

146 **they don't live long enough to do much converting:** A. Dal Bosco et al.,
"Fatty Acid Composition of Meat and Estimated Indices of Lipid Metabolism in Different Poultry Genotypes Reared Under Organic System," *Poultry Science* 91, no. 8 (2012).

146 **If they eat grass or chia:** P. I. Ponte et al., "Improving the Lipid Nutri-
tive Value of Poultry Meat Through the Incorporation of a Dehydrated
Leguminous-Based Forage in the Diet for Broiler Chicks," *Poultry Sci-
ence* 87, no. 8 (2008); P. I. Ponte et al., "Restricting the Intake of a Cereal-
Based Feed in Free-Range-Pastured Poultry: Effects on Performance and
Meat Quality," *Poultry Science* 87, no. 10 (2008); R. Ayerza, W. Coates,
and M. Lauria, "Chia Seed (*Salvia hispanica* L.) as an Omega-3 Fatty Acid
Source for Broilers: Influence on Fatty Acid Composition, Cholesterol and
Fat Content of White and Dark Meats, Growth Performance, and Sensory
Characteristics," *Poultry Science* 81, no. 6 (2002).

146 **If you plunk laying hens on pasture:** H. D. Karsten et al., "Vitamins A, E and Fatty Acid Composition of the Eggs of Caged Hens and Pastured Hens," *Renewable Agriculture and Food Systems* 25, no. 1 (2010).

146 **have more vitamins E and A:** Ibid.

146 **accumulate in a chicken's liver:** C.-Y. Lee et al., "Carotenoid Accumulation and Their Antioxidant Activity in Spent Laying Hens as Affected by Polarity and Feeding Period," *Asian-Australasian Journal of Animal Sciences* 23, no. 6 (2010).

147 **chickens allowed to range on the Tibetan plateau:** T. Sun et al., "Aspects of Lipid Oxidation of Meat from Free-Range Broilers Consuming a Diet Containing Grasshoppers on Alpine Steppe of the Tibetan Plateau," *Poultry Science* 91, no. 1 (2012).

147 **According to Ajinomoto:** The Ajinomoto study refers to glutathione in its reduced form, not its oxidized forInsm. All references to glutathione in this section refer to reduced glutathione.

148 **the result is "chilling injury":** "The predominance of the oxidized state of the tissue as reflected by lower ratios of ascorbate (AA) to dehydroascorbate (DHA) and reduced glutathione (GSH) to oxidized glutathione (GSSG) was linked to the severity of CI symptoms during the last 3–4 weeks of storage," Sukhvinder Pal Singh and Zora Singh, "Postharvest Cold Storage–Induced Oxidative Stress in Japanese Plums (Prunus salicina Lindl. cv. Amber Jewel) in Relation to Harvest Maturity," *Australian Journal of Crop Science* 7, no. 3 (2013).

149 **drenched their stomachs with a terpene in sagebrush:** Luthando E. Dziba, Jeffery O. Hall, and Frederick D. Provenza, "Feeding Behavior of Lambs in Relation to Kinetics of 1,8-Cineole Dosed Intravenously or into the Rumen," *Journal of Chemical Ecology* 32, no. 2 (2006). According to Provenza, in a telephone interview, there was no apparent toxicity during rumen doses. All the sheep did was lick their lips the first time. The second time, they didn't. The toxic effects were most pronounced when the lambs received the terpenes intravenously.

150 **rats that consumed blueberry extract:** A. L. Molan, M. A. Lila, and J. Mawson, "Satiety in Rats Following Blueberry Extract Consumption Induced by Appetite-Suppressing Mechanisms Unrelated to *in Vitro* or *in Vivo* Antioxidant Capacity," *Food Chemistry* 107, no. 3 (2008).

150 **The phenomenon has even been observed by Pancosma:** Related to me by Christian Bruneau of Pancosma, in telephone interview and e-mail correspondence.

150 **Raisins boost the level of a gut hormone:** M. J. Puglisi et al., "Raisins and Walking Alter Appetite Hormones and Plasma Lipids by Modifications in Lipoprotein Metabolism and Up-Regulation of the Low-Density Lipoprotein Receptor," *Metabolism* 58, no. 1 (2009).

150 **eating chili peppers:** M. J. Ludy and R. D. Mattes, "The Effects of Hedoni-

cally Acceptable Red Pepper Doses on Thermogenesis and Appetite," *Physiology and Behavior* 102 (2011); S. Snitker et al., "Effects of Novel Capsinoid Treatment on Fatness and Energy Metabolism in Humans: Possible Pharmacogenetic Implications," *American Journal of Clinical Nutrition* 89, no. 1 (2009); M. Yoshioka et al., "Effects of Red Pepper on Appetite and Energy Intake," *British Journal of Nutrition* 82, no. 2 (1999).

150 **They have only just begun:** John B. Furness et al., "The Gut as a Sensory Organ," *Nature Reviews Gastroenterology & Hepatology* 10, no. 12 (2013).

151 **Injured onion flesh produces compounds:** Emily A. Wilson and Barbara Demmig-Adams, "Antioxidant, Anti-inflammatory, and Antimicrobial Properties of Garlic and Onions," *Nutrition & Food Science* 37, no. 3 (2007).

151 **a flavor compound in nutmeg and parsley:** A. K. Demetriades et al., "Low Cost, High Risk: Accidental Nutmeg Intoxication," *Emergency Medicine Journal* 22 (2005).

152 **the human brain evolved a system:** Edward H. Hagen, Casey J. Roulette, and Roger J. Sullivan, "Explaining Human Recreational Use of 'Pesticides': The Neurotoxin Regulation Model of Substance Use vs. the Hijack Model and Implications for Age and Sex Differences in Drug Consumption," *Frontiers in Psychiatry* 4 (2013).

152 **it causes anxiety and paranoia:** Frank H. Gawin and Herbert D. Kleber, "Abstinence Symptomatology and Psychiatric Diagnosis in Cocaine Abusers: Clinical Observations," *Archives of General Psychiatry* 43, no. 2 (1986).

152 **adds up to about 250 calories:** According to the USDA's National Nutrient Database for Standard Reference, a medium grapefruit is 82 kcal. The calorie figures that follow are also taken from the National Nutrient Database, ndb.nal.usda.gov.

152 **my recommended daily intake:** According to Health Canada, www .hc-sc.gc.ca/fn-an/food-guide-aliment/basics-base/1_1_1-eng.php. I place myself in the "Active Level" group, though even at "Low Active," 250 calories comes in at less than 10 percent.

153 **A veteran magazine publisher I know:** As told to author by Al Zikovitz, president and CEO of Cottage Life Media.

153 **deer, monkeys, or caterpillars:** For more on the meal-limiting effects of plant secondary compounds in primates, see Kenneth E. Glander, "The Impact of Plant Secondary Compounds on Primate Feeding Behavior," *American Journal of Physical Anthropology* 25, Suppl. 3 (1982).

154 **release hormones that trigger satiety:** Sara Janssen et al., "Bitter Taste Receptors and α-Gustducin Regulate the Secretion of Ghrelin with Functional Effects on Food Intake and Gastric Emptying," *Proceedings of the National Academy of Sciences* 108, no. 5 (2011). Inge Depoortere, one of the paper's authors, told me via e-mail that the bitter compound used in the

study was recognized by eight of the twenty-five different bitter receptors in humans.

158 **the lead ingredients are crystalline fructose and cane sugar:** vitamin water.com/files/vitaminwater_NutritionFacts.pdf.

159 **A strawberry contains vitamin C:** This list of micronutrients was assembled with the aid of Donald Davis and various faculty members at University of Florida's Horticultural Sciences Department.

159 **at 4.2:** Nutritional information from www.popchips.ca/popchips/potato /?section=nutrition&flavor=original#.

160 **there are rules that stipulate:** The regulations are taken from Institut national de l'origine et de la qualité, www.inao.gouv.fr. A characteristic description: "Le Mont d'Or est un fromage artisanal. Le lait provient uniquement de vaches de race Montbéliarde ou Simmental Française, nourries exclusivement à partir d'herbe et de foin. Les produits fermentés sont interdits dans l'alimentation du troupeau laitier. Le système d'exploitation est extensif." (Mont d'Or is an artisanal cheese. The milk must only come from Montbéliarde or French Simmental cows, nourished exclusively on pasture and hay. Fermented products are forbidden in the diet. It is an extensive grazing system.)

160 **A modern goat dairy:** I am grateful to the Ontario Ministry of Agriculture and Food and Rural Affairs's Phillip Wilman for information on yields of modern dairy goats.

Eight: The Tomato of Tomorrow

169 **discovered in the late '70s:** The phenomenon of sweet-enhancement through volatiles became explicit in the 1980s, but the additive "cross-modal" interaction between taste and retronasal olfaction is evident in C. Murphy, W. S. Cain, and L. M. Bartoshuk, "Mutual Action of Taste and Olfaction," *Sense Processes* 1, no. 3 (1977). Looking back, Bartoshuk wrote to me by e-mail, "It fascinates me that we were all so slow to take volatile-enhanced sweetness seriously. In those days we looked at volatiles individually and such individual effects are small. The real beauty is the addition across multiple volatiles."

170 **Perhaps most interesting:** The scientific research on tomato flavor and tomato enjoyment portrayed in this chapter is based on interviews, in person and by e-mail and telephone, with Harry Klee, Linda Bartoshuk, and Denise Tieman, all of the University of Florida. Much, but not all, of the scientific ground is covered in Bartoshuk and Klee's "Better Fruits and Vegetables Through Sensory Analysis." See also Klee and Terman's "Genetic Challenges of Flavor Improvement in Tomato," and D. Tieman et al., "The

Chemical Interactions Underlying Tomato Flavor Preferences," *Current Biology* 22, no. 11 (2012).

173 **the twenty most important aromatic compounds in a tomato:** The problem with the old list is that it didn't account for the fact that there can be aromatic compounds in tomatoes at such low levels people aren't able to smell them on their own. For years it was thought these "subthreshold" compounds flew under the radar and, as a result, were ignored. Klee discovered that even at apparently nondetectable amounts, these aromatic compounds contribute to the perception and enjoyment of flavor. The question, of course, is how? There are two possibilities. The first is that, as was said about the inner workings of the nose in chapter 3, a single smell receptor in the nasal cavity can be triggered by more than one compound and, similarly, the same compound can trigger more than one receptor. Subthreshold aromatic compounds, therefore, could tingle some of the same receptors as other compounds in tomatoes and, therefore, rise above the threshold of detection. But there's a more intriguing possibility, which is that the phenomenon is more mental than physical. It could be that a tiny waft of this or that compound is ordinarily ignored by the brain as a kind of olfactory noise, but in the context of a tomato the signal becomes "meaningful." Think of it this way: If you add a violinist to an orchestra, it could improve the music by making it louder—a *physical* phenomenon. But it could also improve the music by making it sound better—a *mental* phenomenon.

Nine: The Gospel According to Real Flavor

181 **Sensory panels laud these potatoes:** W. L. Morris, "Relationships Between Volatile and Non-Volatile Metabolites and Attributes of Processed Potato Flavour," *Phytochemistry* 71 (2010).

182 **lower blood pressure in adults with hypertension:** J. A.. Vinson et al., "High-Antioxidant Potatoes: Acute in Vivo Antioxidant Source and Hypotensive Agent in Humans After Supplementation to Hypertensive Subjects," *Journal of Agriculture and Food Chemistry* 60, no. 27 (2012).

182 **improve inflammation markers:** K. L. Kaspar, "Pigmented Potato Consumption Alters Oxidative Stress and Inflammatory Damage in Men," *Journal of Nutrition* 141, no. 1 (2011).

197 **Label Rouge chickens cost about double:** For information on the chicken market in France, I am grateful to Claude Toudic of Hubbard Breeders.

Appendix: How to Live Long and Eat Flavorfully

201 **Research shows that for infants:** Julie A. Mennella and Gary K. Beauchamp, "The Role of Early Life Experiences in Flavor Perception and Delight," in Dubé, *Obesity Prevention.*

201 **and cattle, too:** R. W. Wiedmeier et al., "Eating a High Fiber Diet During Pregnancy Increases Intake and Digestibility of a High Fiber Diet by Offspring in Cattle," *Animal Feed Science and Technology* 177 (2012).

202 **Moms who eat junk food:** T. M. Cutting et al., "Like Mother, Like Daughter: Familial Patterns of Overweight Are Mediated by Mothers' Dietary Disinhibition," *American Journal of Clinical Nutrition* 69, no. 4 (1999); J. R. Gugusheff, Z. Y. Ong, and B. S. Muhlhausler, "A Maternal 'Junk-Food' Diet Reduces Sensitivity to the Opioid Antagonist Naloxone in Offspring Postweaning," *FASEB Journal* 27, no. 3 (2013).

Index

Page numbers beginning with 223 refer to endnotes.

About the Author

Mark Schatzker is an award-winning food journalist and the author of *Steak: One Man's Search for the World's Tastiest Piece of Beef*. His work has appeared in *The New York Times, The Wall Street Journal, Condé Nast Traveler,* and *Best American Travel Writing*. A radio columnist for the Canadian Broadcasting Corporation, Schatzker also writes frequently for *The Globe and Mail* and *Bloomberg Pursuits*. Visit him online at www.markschatzker.com.